Evolution-Development

And

The Master Development Program Hypothesis

Overview 2016 with Updates

George E. Parris

Dedication

"Hypotheses are not statements of truth, but instruments to be used in the ascertainment of truth. Their value does not depend upon ultimate verification, but is to be measured by their effect upon scientific research."

...C. Stuart Gager, University of Missouri 1909 (Translator's notes on *Intracellular Pangenesis*, 1910)

None of this work would have ever seen the light of day without the open-minded and fearless work of Dr. Bruce Charlton and Dr. William Bains, editors of the journals in which the original work was published.

Preface

Since the publication of the first edition of this book on Amazon/Kindle (April 8, 2013) the medium has improved to make much better use of graphical material that is very useful in scientific articles. In addition, in the last seven years since the publication of the idea and the three years since the first edition was published, academic research has continued to move forward at a blinding pace producing additional insights without, to my knowledge, contradicting any of the primary elements of the Master Development Hypothesis. This hypothesis was originally published in several parts including the following:

Parris GE. 2009. A Hypothetical Master Development Program for Multi-cellular Organisms: Ontogeny and Phylogeny. 2009. *Biosciences Hypotheses.* 2:3-12

Parris GE. 2010. Developmental diseases and the hypothetical Master Development Program. *Med. Hypotheses.* 74(3):564-73.

Parris GE. 2010. Scope of Medical Implications of the Master Development Program. *Med. Hypotheses.* 74(5):953.

Parris GE. 2011. Asymmetric Division and the Immortal Strand Hypothesis. *Hypotheses in the Life Sciences.* 1(2):52-55.

Parris GE. 2011. The Hopeful Monster Finds and Mate and Founds a New Species. *Hypotheses in the Life Sciences.* 1(2):1-6.

Parris GE. 2013. Application of a Hypothesis to Speciation in Hominidae. *Hypotheses in the Life Sciences.* 3(1).

Parris GE. 2014. The Developmental Hourglass. Self-published on Amazon/Kindle.

At this point, I do not see the need to include discussions of how RNA transcripts regulate transcription of genes (selecting, activating, de-activation) or affect post-transcriptional activities that determine the proteasome so I have not included that discussion in this draft. Three years ago, I felt I had to justify those ideas; now I do not.

During the frenzy for the last couple of years of the Human Genome Project (circa 2000), my interest turned to how development of complex asymmetrical organism could reproducibly follow form a single cell. It had to be a matter of chemistry. My examination of the literature told me that I was interested in the new field called of *evolutionary developmental biology* (a.k.a., Evo-Devo). Apparently, I was not the only person interested in this problem, but I think I am the only chemist.

To biologists, the single-cell (fertilized egg) seemed to need an outside guidance (chemical gradients, gravity, electromagnetic fields, thermal gradients, etc.). A check of the textbooks and then current papers, pointed to a theory (originally proposed by Lewis Wolpert [1-4]) of multiple chemical gradients guiding activation of receptors on various cells. One of the greatest proponents of this idea seemed to be Sean B. Carroll who discussed and endorsed this theory in his book *Endless Forms Most Beautiful* (W.W. Norton, 2005).

But, my experience as a chemist suggested that such gradients are unreliable, imprecise (on the cellular scale) and unstable. Such gradients would be affected by physical motion, thermal gradients, hydrodynamic pressure and a wide variety of other factors. Looking at the precision and reproducibility of the development program as demonstrated by development of identical twins, the fact that human anatomy can be charted in detail (reproducible across billions of individuals) and the success of humans and other species under a wide variety of climatic, cultural and dietary conditions told me that the development program must be hardwired and extremely reliable. Perhaps these same considerations caused Wolpert to distance himself from the gradient theory after 2009 [5-7]. But, it is still the basic model used to explain development [5, 8].

Of course, no one had any relevant biochemical data at the time (1998-2000), so I began thinking about how I would design such a system using what I knew about the genome and DNA. I worked on this idea for several years and eventually tried to find a journal that would consider the work. I was unsuccessful until a new hypothesis journal was started by Elsevier [9]. Unfortunately, that journal was terminated within the first year. This was a disaster since they held the copyright to the work and the journal was not indexed in *pubmed* (I believe it is in *Science Direct*). I managed to convince the editor of *Medical Hypotheses* (Elsevier) to publish an updated version of the idea with a slant towards medical applications [10, 11]. One of the most important concepts in the hypothesis is the idea about how a cell of one phenotype (proteome) can differentiate to produce a cell with a different phenotype (proteome). I had described that in the first publication [9], which was not widely indexed; but not in the second publication [10], which was widely indexed. Thus, I elaborated this idea and published it in yet another hypothesis journal *Hypotheses in the Life Sciences* [12], which has also stopped publishing, but at least the articles are supported by a university on the internet. Moreover, I have published key elements via Amazon-Kindle.

It has been several years since these papers were published and experimental data in relevant fields continues to pile up in the peer-reviewed literature. *I have not seen anything that contradicts the hypothesis as originally proposed*, but no one to my knowledge is actively trying to test the hypothesis. Thus, the first objective of this manuscript is to summarize the model as it has been proposed [9, 10, 12] with recent citations where appropriate. I also want to show that the hypothesis is consistent with what is known about differentiation, evolution and speciation.

In the first edition, I made an effort to include general information about the genome for casual readers. This edition is targeted at readers with at least a basic knowledge of cell biology.

George E. Parris
Gaithersburg, Maryland, USA
December 2015
Updated August 2019

Conventions

There seems to be a variety of ways to distinguish codes in DNA, RNA transcripts of those codes and proteins. Therefore, I have decided to follow the following conventions:

DNA codes (i.e., genes) in italics

RNA transcripts in regular script

Proteins (peptides, etc.) in ALL CAPS. However, there are many exceptions especially for well-established names and identifiers with numbers (e.g., Bcl-2).

For logistical reasons, (i.e., my subscription to EndNotes as lapsed) I am including new (post-2015) and recently discovered earlier citations in this edition as footnotes.

Introduction

Genes and Pangenes

Charles Darwin and Gregor Mendel are given credit for the concept of heredity as we know it. The history is much more complex with contributions from a variety of people. But, in the interest of simplicity, I'll focus on Darwin. Long after publishing his book on evolution, Darwin published a hypothesis about the mechanism through which traits were inherited [13-16]. He called the physical objects that carried the traits "pangenes." Pangene was later shortened to "gene" by Hugo de Vries [17-19] in the early 1900s. So far; so good. But, when A. H. Sturtevant [20] a student of T.H. Morgan published a paper entitled "The Spatial Relationship of Genes" in 1919, our idea of a gene was changed from a *carrier of a trait* to a *sequence of DNA that codes for a specific protein.*

In some cases, loss or mutation of a specific set of codes resulting in the loss of a protein does produce an observable change in a trait of a complex organism (Morgan's group studied fruit flies). For example, loss of the codes for the enzymes that produce the pigment melanin produces albino mammals. But normal mammals are not just black and colorless. The variety of colorations found in mammals is caused not by the loss or gain of a set of codes for a specific protein, but rather by changes in the development of the organism (i.e., when, where and for how long those genes are expressed).

Thus, Darwin's "pangene" concept is much more tied to some code regulating differentiation/development than it is to a specific protein. I have recommended reintroduction of the term "pangene" to refer to codes for developmental processes, but what has happened is that we now use the term "protein-coding gene" to refer to those codes in DNA that actually code for a protein. If the hypothesis that I have developed is found to be correct, "pangene" may reenter the biology texts.

Chemistry is the Basis of Biology

If biology works, it works because the underlying chemistry works. When I began my inquiry circa 2000, I knew that there must be sound chemistry that explains biological development. Thus, disregarding the hypotheses accepted by biologists, I took the position that I should start from the ground up and invent a system that had the necessary characteristics to explain development. However, it had to be consistent with well-established biological observations (but not necessarily the interpretations that biologists gave to those observations). I took the position that biology *on this planet* may not work as I have described, but it might work that way somewhere in the universe.[1]

I will describe the model that came from that exercise below. Once I had built the *development* model, I was surprised and rewarded to realize that my model could also explain *evolution* of morphology, biochemistry and even behavior. Indeed, it dawned on me that it *is essential* that the true model for *development* must also explain *evolution*. This realization, caused me to have more faith in the model, which I began calling the Master Development Program.

There were a number of problems that I faced in creating the model. I described these in some depth in my first publications [9, 10]. Here I will not spend so much time describing how I discovered the problems but rather focus on how I solved them.

---

[1] In the future I plan to write a book on abiogenesis. It is my opinion that life throughout the universe has followed a very similar pattern with identical chemical building blocks.

The Fundamental Problems in Modeling Development

The development of complex, multi-tissue, asymmetric organisms from a single (nominally spherically symmetrical) cell presents a number of fundamental problems. Among these are as follows:

(i) How transcription of the genome is regulated;

(ii) How a single cell can not only change its transcriptome but (*a*) do it in a particular sequence, (*b*) branch to form numerous clones with different transcriptomes, (*c*) and at each step proliferate for an exact number of cycles needed to produce the ultimate (adult) morphology, and (d) do all this reliably;

(iii) The information needed to accomplish the differentiation must be stored in the genome, but how and where are not obvious;

(iv) The mechanism of storage of the information must be consistent with what is known about (*a*) sexual reproduction, (*b*) developmental diseases (e.g., the effects of teratogens), and (*c*) ontogeny;

(v) The mechanism of storage must be able to account for orderly evolution of the species consistent with known phylogeny; and

(vi) The mechanism cannot be contrary to well established biological observations (although it may be contrary to conventional interpretations).

In developing a hypothetical system, in principle, I did not need to consider biology on this planet (hence, my ignorance of biology was not a problem). But, in practice, it was clear that (i) evolution on earth had already solved many of the problems and could be relied upon for guidance and (ii) if the system that was derived happened to be the system used in the biological system familiar to us on this planet, the result would be of substantial practical utility in all fields of biology. Thus, I considered these criteria as primary biological restrictions on the hypothetical mechanism I wished to devise from chemical principles. Of course, the chemistry must be plausible as well.

It soon became clear that my problem of development was primarily associated with the process of differentiation of cell clones. In particular, I chose the term "program" very intentionally because I wanted to make sure to distinguish the "program" (i.e., the sequence by which cells differentiate and the organism is developed, ontogeny) from the idea of a "plan" that merely shows the end result. As a matter of fact, I envisioned the Master Development Program much like a computer program that calls out subroutines (transcription of proteins) as it is executed. But, a simple computer-like code is not complete, because the Master Development Program needed to be able to sequence in such a way, that it branched repeatedly (i.e., differentiated) producing branches that operated concurrently in different cell clones to produce cells of completely different types. And, coordinate the differentiation and expansion of all cell clones.

The Difference between Plants and Animals

For simplicity let me distinguish two types of development: Plants and animals. Plants, in my view, are indeed driven primarily by simple gradients provided primarily by moisture, sunlight and gravity. You will find plants species defined by the types of leaves, bark, branching, root structure and habit (tall and straight or short and bushy, etc.). However, you will *not* find detailed anatomical charts of different species of tree or other plant because every plant is different.[2] Even cloned

---

[2] It is my observation that the leaves of some species of trees have very similar patterns of veins and stems, which to me suggest a program.

plants branch differently (following similar patterns) in the same environment. Plants regenerate readily. If Wolpert's hypotheses [2, 3] were limited to plants, I would believe that they might be correct.

Individual animals, on the other hand, develop exactly true to their species pattern, right down to minute nerves and blood vessels.[3] Moreover, the development is clearly independent of general environmental conditions, diet, etc. Even teratomata and fetuses-in-fetu develop as though the cells are programmed internally... not driven externally. While the teratomata are generally much less organized, they contain fully developed elements in many cases. Their reputation for being more likely to produce dangerous tumors can be explained by retention of some high-level stem cells while fetus-in-fetu seem to have progress more evenly in all regards; otherwise there seems to be little difference. [21]

I hypothesize that the evolution of complex animals required the acquisition of a system for recording a program that guides development. This argument implies that plants have a much less specific program (or perhaps no real program at all). For example, in each leaf of an oak tree, the veins follow a similar pattern, but the patterns are not identical. In contrast, the blood vessels

---

[3] There are a few possible variations, but these appear to be hereditary, not random.

in each person's hands are identical and the left and right hands are mirror images.

Thus, what I am really looking for is the system that animals use to record the program (i.e., how evolution occurs) and how the program is executed (i.e., how animals develop).

I recently discover the work of the late Dr. AJ Klar. He worked on some of the same problems and starting with simple systems and by the early 2000s was starting to think about development of higher organisms.[4] In his 2016 paper, came to the same conclusion that I had: "We also argue against the conventionally invoked morphogen model of development." And the model that he came to for asymmetric cell divisions (in 2016) is closely related to the model I published (in 2009, 2010 and 2011).

---

[4] Armakolas A, Klar AJ. 2006. Cell type regulates selective segregation of mouse chromosome 7 DNA strands in mitosis. *Science.* 311(5764):1146-9.

Klar AJ. 2008. Support for the selective chromatid segregation hypothesis advanced for the mechanism of left-right body axis development in mice. *Breast Dis.* 29:47-56.

Klar AJ. 2016. Split hand/foot malformation genetics supports the chromosome 7 copy segregation mechanism for human limb development. *Philos Trans R Soc Lond B Biol Sci.* 371(1710).

Part 1. Overview of the Master Development Program

1.1 Introduction

In the original papers (2009-2011), I laid out the logic of my analysis and led the reader to my conclusions. In Part 1, I will present an abstract of the model and then explain its evolution and functions in more detail in other Parts.

What is Development of an Organism?

My interpretation of *development* is that a single cell (zygote, fertilized oocyte) systematically undergoes sequences of proliferation and differentiations that are defined by successive generations of the clone and sub-clones. In each generation a cell has three choices: (i) reproduce itself exactly through mitosis (i.e., proliferate), (ii) produced two identical *new* cells with new transcriptomes and proteasomes (i.e., simple differentiation) (iii) produce a new type of cell while maintaining the original cell type (i.e., founding a new clone; the cells that can do this are called *stem cells*).

Stem cells produce "forks" in the branching family of cells.[5]

Based on their behavior the cells are labelled as (i) stem cells, (ii) progenitor cells, or (iii) terminally (i.e., fully) differentiated cells.

> **Stem cells** can undergo simple proliferation (making more of themselves); or a stem cell can produce a differentiated cell (which may be a lower order stem cell or a progenitor cell) and a new copy of itself through *asymmetric division* (which will be discussed in some detail below). We speak of totipotent (omnipotent) and poly-potent stem cells. Obviously, the earlier a stem cell occurs in the sequence of differentiation, its ability to recapitulate more specialized clones is greater.

> **Progenitor cells** give rise to new types of cells as they proliferate. Some generations of these cells may be low-order stem cells.

> **Terminally differentiated cells** only reproduce themselves via proliferation (mitosis).

---

[5] Of course, for sexual reproduction, meiosis is required as well.

Development of an Organism

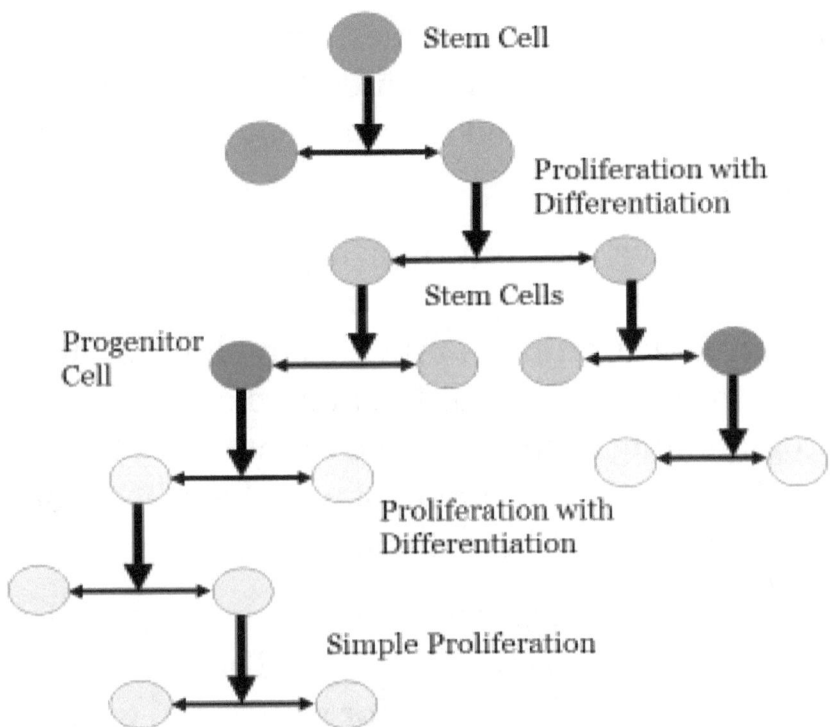

Stem Cell

Proliferation with Differentiation

Stem Cells

Progenitor Cell

Proliferation with Differentiation

Simple Proliferation

Terminally Differentiated Cells

After the initial burst of differentiation that forms the principal body parts, stem cells tend to go into a state of suspended animation (not senescence), and reproduce themselves rarely. In this way, they avoid damage to their genome, and they standby to produce new cells for the replacement of destroyed clones. Progenitor cells may simply proliferate or may proliferate with differentiation. Terminally differentiated cells are only able to reproduce themselves.

Reprogramming (i.e., returning a cell to an earlier point in the differentiation program) can be accomplished. The most dramatic example is provided by cloning entire animals from terminally differentiated cells. Reprogramming of course, is a current topic of great medical interest and a variety of ways have been found to erase the marks of differentiation.

Thus, development is clearly "programmed" and clearly the program evolved (step-by-step). The notion of "random mutations" of genes (or pangenes), which might occur anywhere in the program seems to me to be absurd. Imagine a computer program with many lines of code. If you randomly change a line of code, you would have to make many other changes in the program to make it work.

There is simply not enough *evolutionary power* in higher organisms for purely random modifications in the development program to produce productive change. By "evolutionary power" I mean the opportunities for testing and rejection or acceptance of modifications. In single-cell organisms (population extraordinarily high; with a new generation of cells every few hours; and time to evolve (billions of years)) there was enough time to evolve functional proteins and genomes by purely random mutations.

Number of bacteria in top 10 m of water on earth:

$(10^4$ cells/cm^3) x $(4$ x 10^8 km^2) x $(10^{10}$ cm^2/km^2) x $(10^3$ cm$)$ = 4 x 10^{25}

Frequency of reproduction: 2/day

Period of evolution to produce first multi-cell organisms: 10^9 years

Evolutionary power:

4 x 10^{25} x 2/day x 365 days/year =

3 x 10^{28} events/year

Total Evolutionary potential of bacteria:

4 x 10^{25} x 2/day x 365 days/year x 10^9 years =

3 x 10^{37} events

The reason that higher organisms use virtually the same proteins as bacteria and these same medium of inheritance (i.e., DNA) is that higher organisms have not had (and *will never have*) enough time to make any fundamental *improvements in either proteins or DNA*. For example, the *GLO* gene (L-gulono-γ-lactone oxidase)[22] in most primates was damaged about 61 million years ago (MYA). In spite of the fact that this is conditionally lethal gene, it has not been repaired or replaced (which would require a sequence of favorable random mutations).

If you compare the figures in the above box to higher organisms, you will see why random mutations do not improve higher species…although random mutations can degrade proteins in higher species leading to evolutionary dead-ends.

The only way to isolate an anatomical (or biochemical feature) and test it in a complex organism is through *terminal addition*. Specifically, terminal addition to the development program [9, 10]. Moreover, to ensure that advances spread through a population, the species must have explicit sexual reproduction.[6]

---

[6] Crossing-over in meiosis is an additional feature, which helps identify evolutionary improvements.

1.2 Evolution of the Master Development Program

Where is the Master Development Program?

I hypothesize that there is a Master Development Program encoded in the genome and that the most important parts are encoded in the pericentromeric heterochromatin. (There is indication that sub-telomeric regions and even dispersed regions of heterochromatin may play similar roles in some species. [23])

If species-specific morphology is controlled by the Master Development Program putatively located in the pericentromeric heterochromatin, then one would expect that this part of the genome (more than the euchromatin or protein-coding genes) would show systematic variation across related species. Indeed, two types of variation are observed: (i) Pericentric inversions (with minimal disruption of the coded sequence) appear to be the genomic feature separating species (see below). (ii) The organization of the pericentromeric region should show some correlation with the phenotype of the species. (It is noteworthy that the number and types of genes do not seem to have much effect on the phenotype... Compare humans and chimps; same protein-coding genes, different phenotypes.) In contrast, it has been observed that pericentromeric sequences generally vary with species (corresponding to different phenotypes) [24-43].

How are Species Established?

Alfred Russel Wallace [44] and Charles Darwin independently deduced that the mechanism of *morphological evolution* of organisms can be explained by *natural selection*. In his haste to establish a broader priority on the idea, Darwin notoriously self-published a document known widely as **On the origin of species** *by means of natural selection or the preservation of favoured races in the struggle of life* (Murray, London, 1859). In this document, he claimed that *evolution creates species*; and (against the advice of Thomas Henry Huxley) Darwin insisted that speciation can only happen gradually [45].

After the discovery of protein-coding genes in the chromosomes in 1920 [20, 46], the assumption adopted by most evolutionary biologists was that evolution (e.g., change in phenotype) must involve mutations in the protein-coding genes. This concept has persisted, although it is now known that the protein-coding genes and the machinery that makes them work is available in simple bacteria; and these biologists continue to marvel at the similarity of human and chimp protein-coding genes.

As discussed above, there simply is not enough evolutionary power in higher organisms to create new protein-coding genes and random mutations are far more likely to destroy genes than improve them. Nonetheless, mainstream biology still seems to cling to this idea although the emphasis has shifted to the non-coding elements in the genome [47, 48]. Mattick [48] references "feed-forward RNA regulatory networks" which seem to be spread through the genome as the system modulated by mutations leading to evolution (i.e., systematic differentiation). But, here again random mutations would appear to me to be non-productive or counter-productive.

It is academically dangerous to challenge Darwin. Nonetheless, Richard Goldschmidt noted that speciation does not require a gradual process and I have added that new species need not vary significantly from the parent species [45, 49]. Several generations of biology students (1945-2005) have been treated to classroom jokes about Goldschmidt's "hopeful monster" dilemma [50]. The joke being that if there were a sudden mutation that created a "new species" (i.e., dramatically new morphology) how could this "monster" find a mate to extend the species.

There is a simple and (at this point in time) a fairly obvious solution to Goldschmidt's "hopeful monster" dilemma. Although the protein coding genes of similar species are almost identical, it is well known that the most obvious differences among closely related species (e.g., the primates) is the existence of pericentric inversions. Keep in mind that within each species the inversions are homogeneous (i.e., on both alleles of the respective chromosome) and that each species is fertile within its species (but not fertile outside its species) [51, 52]. It is also known that pericentric inversions are relatively common (even in humans [53-55]) and that they interfere with reproductive success (i.e., when an inverted chromosome and a chromosome of normal parity are present, the fetus may not develop). In particular, individuals with *homogeneous inverted parity* can be produced by inbreeding (e.g., among *heterozygous inverted* parity cousins). Then *homozygous inverted* parity individuals cannot breed successfully with morphologically identical cousins with *homozygous normal parity.*

Nonetheless, breeding is successful with individuals that share the pericentric inversions [45, 56-58]. I believe this to be the solution to Richard Goldschmidt's "hopeful monster" dilemma. This mode of speciation is consist with the evolution[7] and operation of the Master

---

[7] García-Souto D, Pérez-García C, Pasantes JJ. 2017. Are Pericentric Inversions Reorganizing Wedge Shell Genomes? Genes (Basel). 8(12). pii: E370.

Development Program proposed below. I have pointed out that although we and our close primate relatives obviously reproduce readily *within* our species, inter-species hybrids are prevented by multiple pericentric inversions. Indeed, I use the pericentric inversions to mark the number of distinct species separating us and our close cousins from our common ancestor [45, 49, 59]. This should be a useful tool for paleontologists trying to sort out species from bones.

How does the Master Development Program Evolve?

Although I agree with John S. Mattick [60-63] that regulatory RNA is the primary messenger of differentiation, I believe that systematic evolution and differentiation requires a compact, linear program that evolves primarily by terminal addition. Terminal addition can explain two important elements of evolution: heterochrony [64-66] and punctuated evolution [67-69]. The primary code evolved (and continues to evolve) by terminal addition (insertion) of retro-transcripts of elements from (retro) viruses [70-72] and non-coding elements from gene transcripts (e.g., initiation sequences and introns) *at the centromere* [73].

Evolution of the
Hypothetical Master Development Program

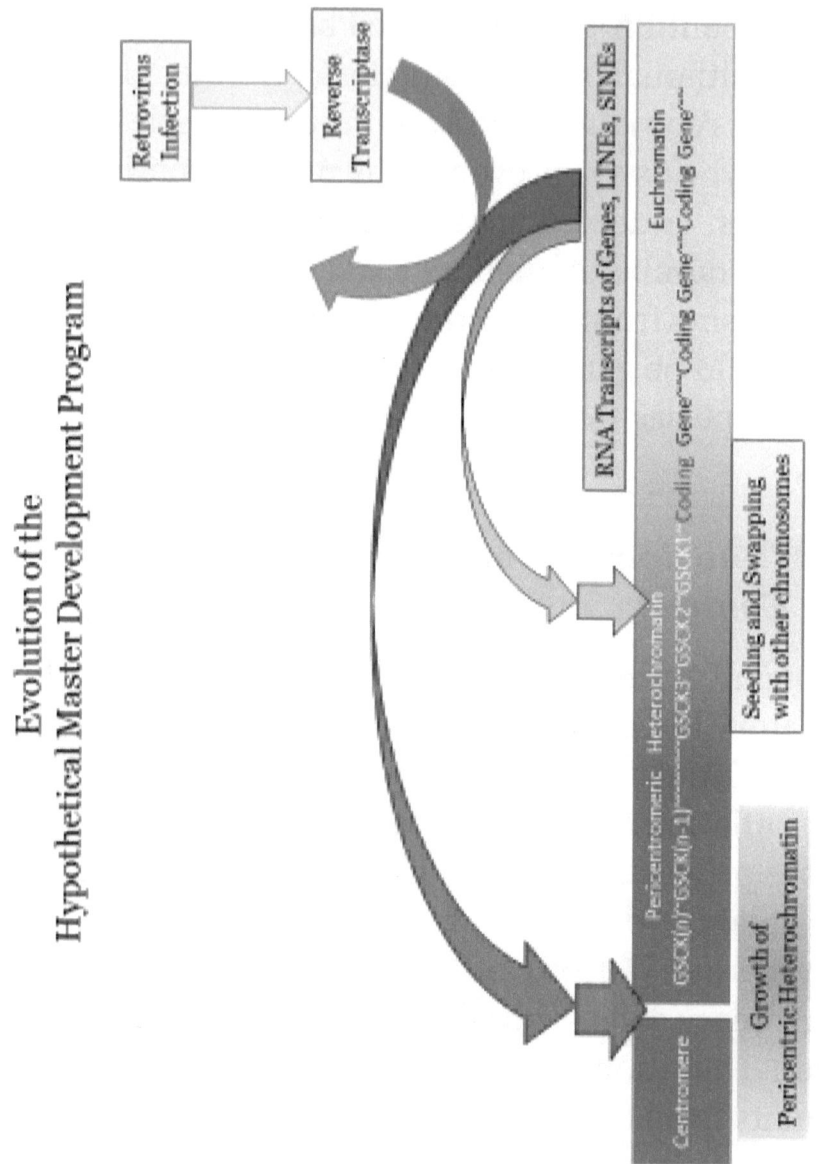

Thus, the most recent (latest addition) part of the code is nearest the centromere and the oldest elements that were added first are located at the junction of the pericentric heterochromatin and the euchromatin. This arrangement has been demonstrated experimentally by Alexandrov et al. [74-78] and Schueler et al. [79-82].

Nonetheless, the distal pericentric heterochromatin (i.e., the oldest parts of the code) is subject to random mutations because of frequent breakage and insertions (seeding and swapping) at the transition from euchromatin to heterochromatin as shown by Horvath et al. [40, 72, 83-87]. The figure above summarizes the evolution of the MDP.

Retroviral Pandemics

The prominent role of reverse transcriptase in the evolutionary scheme shown above suggests that evolution of the MDP would be fastest during periods of retroviral pandemics [88-92]. Indeed, the genomes of all higher species are peppered with elements derived from retroviruses [70-72, 93]. Endogenous retrovirus (ERV) elements *that have survived in genomes* are frequently located in the pericentromeric heterochromatin and promoter regions of protein-coding genes. The point here is that except for the breakpoints associated with the pericentromeric heterochromatin (associated with evolution of the MDP, see above), ERVs are likely inserted randomly into the genome. Some of these, of course, cause lethality or sterility of the offspring. The fact that the ERV-derived elements have survived in promoter regions of protein-coding genes is strong evidence that they have improved the fitness of the species that carry them (i.e., positive selection) [36, 39, 40, 88, 89, 94-112]. The inclusion (and survival) of ERV-elements in the pericentromeric heterochromatin and in the promoter regions of protein-coding genes ensures that *nuclear messenger RNA* (nmRNA) transcripts from the pericentromeric region are complementary to the promoters of active genes. The same argument can be made for elements inserted into the heterochromatin that are derived from transcription of protein-coding genes (e.g., introns) [100].

Coevolution of the
Adult and the Embryonic Forms

It is implicit in the organization of the pericentromeric heterochromatin (as discussed above) that execution of the MDP begins at the point most distant from the centromere and moves progressively towards the centromere as each generation of the cell clones is produced. This order of execution fortuitously opens both the initial stages of development and the final stages of development for evolution (i.e., simultaneous modification). This is important because *as the adult form is modified the embryonic form must also be modified* (i.e., the adult and embryo/fetus must be compatible for the species to survive).

Darwinists including Ernst Haeckel [113-115] did not anticipate this dilemma and made the assumption that development and evolution would be linearly related (like the straight arrows in the figure below). They assumed that all the adult forms would converge backward to a point and that each species would be characterized by a separate phylotypic stage.

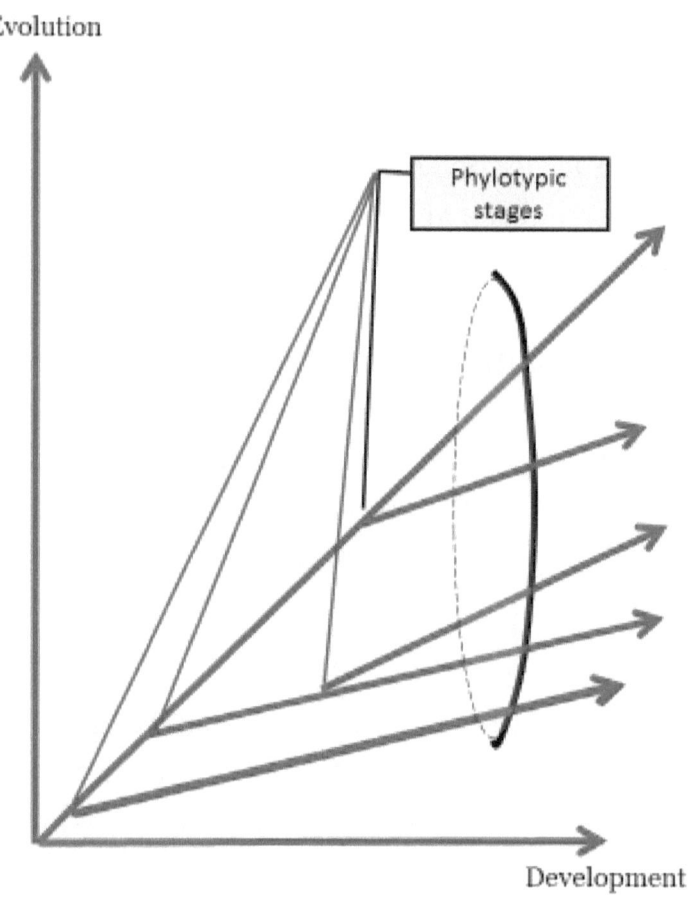

However, once more detailed data were available, it was apparent that the embryonic forms actually converged to the apparent phylotypic stage and then radiated outward as shown below. This behavior has caused consternation among the Darwinists who cannot account for the observations with random mutations [116-120]. However, it is explained by the MDP hypothesis.

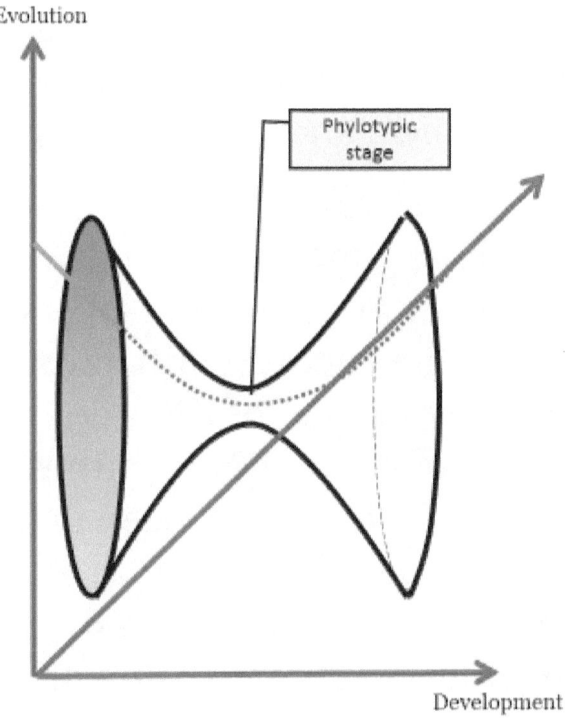

How is the Master Development Program Conserved?

Orderly evolution of the MDP by terminal addition at the centromere with corresponding modifications at the junction with the euchromatin is essential. Any sort of randomization of the code between the start and the finish would result in chaos. Thus, incorporation of the code within the pericentromeric heterochromatin protects it from random (mechanical) breaks and the action of mutagens (chemical and radiological).

However, the most vulnerable time for the MDP code may be during meiosis when the chromosome is exposed to cross-overs with its homologue. In the euchromatin, crossovers appear to be a byproduct of DNA double and single strand breaks (following DNA synthesis) as the cell enters prophase I. There has been debate whether crossing-over causes close synapse formation or synapse formation facilitates crossing-over. There is also some uncertainty regarding proximity of the chromatids during the process. I'm bettering that, when these issues are sorted out, there will be a role for non-coding RNA. Regardless, there is evidence that specific proteins help protect the pericentromeric DNA from crossing-over during meiosis [121, 122], which might disrupt the essential MDP code.

1.3 Operation of the Master Development Program

It is now well-known that the pericentromeric heterochromatin is transcribed [103, 123, 124]. Transcription of the pericentromeric regions is coordinated with the cell cycle. Specifically, it has been shown that pericentromeric regions are transcribed [110, 125]: (i) late in the s-phase (just before G_2) where it appears to be critical for epigenetic marking of chromosomes and chromocenter formation at the two-cell stage [126-129] and (ii) at the beginning of G_0 phase where it appears to initiate the appropriate transcription for that generation of the clone [130-135]. It is further been shown that important developmental events seem to be timed by elements in the pericentromeric heterochromatin [136, 137].

Model of Genome Regulation

Prokaryotes have no nuclear membrane and genome regulation is achieved by placing genes that code for proteins needed for several steps in a metabolic process into a (localized) operon that is activated directly by interactions of metabolites and nutrients with DNA binding proteins that initiate transcription. This mechanism ensures that the relevant proteins are synthesized at the same time and in the correct relative amounts to facilitate the necessary biochemical transformations inside the cell. For example, in the following metabolic process proteins (enzymes) A, B and C are needed to convert the Nutrient into the Metabolite:

Nutrient →[A]→ Intermediate 1 →[B]→ Intermediate 2 →[C]→ Metabolite

If enzyme "A" is present but enzyme "B" is missing, "Intermediate 1" would accumulate and no "Metabolite" would be produced. Suppose that "Intermediate 1" is toxic to the cell. Obviously, all three enzymes are required at the same time. So, by physically linking the protein-coding genes (*A, B, C*) together in the circular chromosome under a single regulatory unit, the cell always makes the same enzymes at the same time and in the correct proportions. The simple logic is that if there is a high concentration of the nutrient and none of the metabolite, then the operon (*ABC*) is activated. See, for example, the *lac* operon [138, 139].

In eukaryotes, however, the genome is larger and split into numerous chromosomes. Genes that code for proteins that are needed for specific metabolic functions are not co-located. These facts were known in 2000 and led me to the immediate conclusion that the "delocalized operons" of eukaryotes required a mode of regulation that could reach them remotely (i.e., transactivation). The obvious answer to this problem was to invoke various sorts of *nuclear messenger RNAs* (nmRNAs) with appropriate sequences to target the genes of interest. Some of these nmRNAs could be generated by excision of the introns from the gene transcripts themselves producing a cascade of gene activation. However, the initiation of the process *in a specific generation of the cell clone* would reasonably involve a long non-coding RNA (lncRNA) cut into appropriate activating and deactivating nmRNAs. This idea was virtually unknown in 2000 [140], but has now blossomed into routine knowledge [141-143]. This concept formed the heart of my Master Development Program hypothesis.

There has been an upsurge in interest in circular RNAs derived from introns or introns-and-exons in the last few years [142, 144-147]. These structures seem to arise from lariats formed during splicing and they have been found to have roles in regulation of gene transcription [142].

It is hypothesized that the pericentromeric heterochromatin is subdivided as Generation Specific Control Keys (GSCKs). A GSCK is transcribed into a long non-coding RNA (lncRNA) in each generation of the cell. The GSCK is then cut into a variety of individual Gene Control Keys (GCKs) that act to either activate or suppress specific genes.[8] Most of these RNAs are used as nuclear messenger RNAs (nmRNAs) that fan out in the genome and locate specific genes (on numerous chromosomes) primarily by complementary matching the genes' promoters, introns or untranslated regions (UTRs).[9] Obviously, these GSKs (collectively "the GSCK nmRNAs") thus determine the transcriptome of the cell *in that particular generation* of the clone. Some of the fragments of the GSCK may be modified and dispatched out of the nucleus as coding or non-coding mRNAs. (For example, some extra-nuclear non-coding RNAs may be targeted to the mitochondria to regulate their genomes or even released from the cell in exosomes to influence nearby cells. [148-160])

---

[8] When I first developed this idea in the early 2000s, the role of ncRNA was little known and all of this was speculative. Now, it is well-established.

[9] Presumably, GSKs targeted to introns and UTRs facilitate specific isoform transcriptions of these genes.

The cutting of lncRNAs (transcripts of the GSCK) is accomplished by various RNase IIIs especially DROSHA/DICER-like [161-175] with ARGONAUTE-type/PIWI-type proteins. The effects of nmRNAs include activation of transcription [107, 130, 176-179], suppression of transcription, suppression of transcripts (RNAi), splicing transcripts, et al. These frequently involve Oct and Nanog genes associated with stem cells [134, 180].

Ultimately differentiation of cell clones is accomplished by "semi-permanent" methylation and demethylation of the CpG units in promoters for individual protein-coding genes. As one would expect, failure to accomplish this task, leads to developmental abnormalities.[10] The details of these process are beyond the scope of this discussion as they are merely the tools used by the Master Development Program to implement transcription or repression of protein-coding genes.

---

[10] Jin B, Tao Q, Peng J, Soo HM, Wu W, Ying J, Fields CR, Delmas AL, Liu X, Qiu J, Robertson KD. 2008. DNA methyltransferase 3B (DNMT3B) mutations in ICF syndrome lead to altered epigenetic modifications and aberrant expression of genes regulating development, neurogenesis and immune function. *Hum Mol Genet.* 17(5):690-709.

Operation of the
Hypothetical Master Development Program

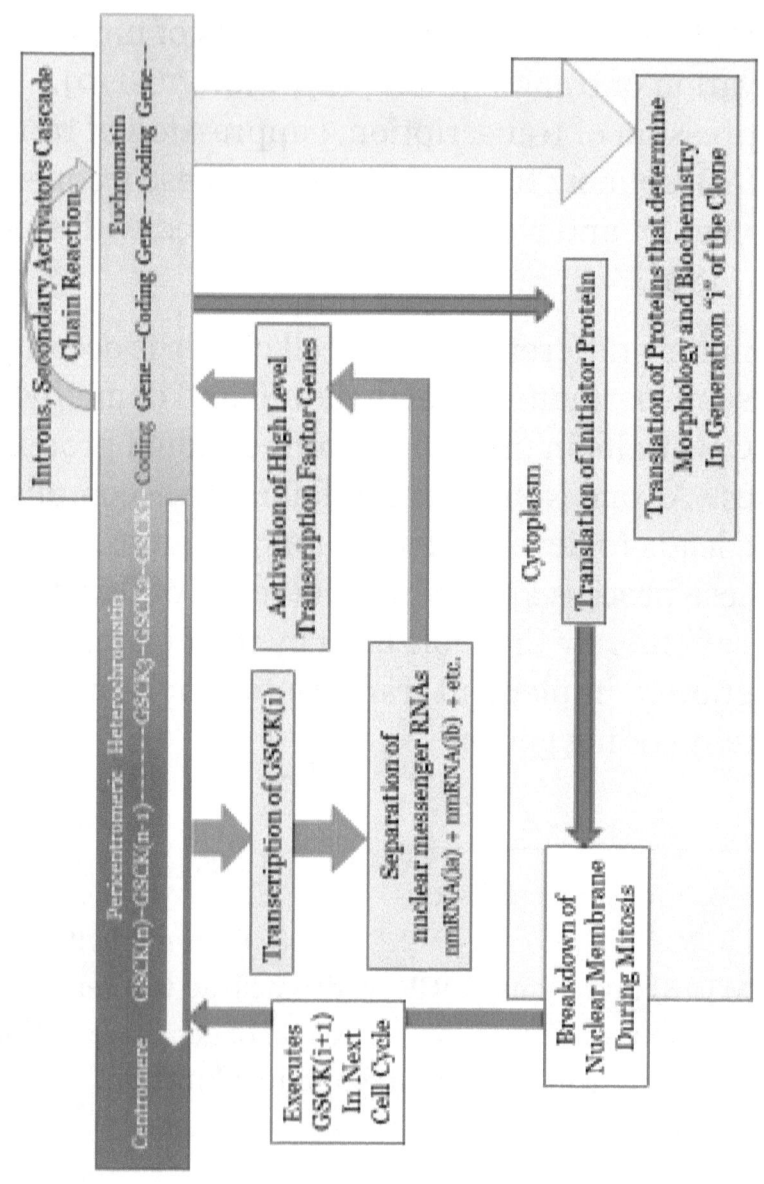

In the diagram above, the white arrow indicates the direction of execution of the MDP encoded in the pericentric heterochromatin. The blue arrows trace the processing of the GSCK (for the ith generation) into GCK nmRNAs, which activate or suppress specific genes (the details of these activities are discussed below). Once protein-coding genes (and/or intergenic non-coding sequences) are transcribed they begin a cascade (chain reaction) of transcription throughout the genome. Thus, the generation-specific transcriptome is produced (light salmon-colored arrows).

The major problem in making this system work is the ability to transcribe the MDP in steps corresponding to the generations of the cell clone. This process follows the red arrows. The basic idea is straightforward and was first demonstrated by Blow and Laskey in 1988 [181].

Replication Licensing and Transcription Initiation

Since 1988, J.J. Blow has published well-over 100 abstracted articles that relate to the regulation of replication in the cell cycle [182-184]. The basic concept is that the principal inhibitor of the *replication licensing factor* (i.e., geminin, GMNN) is presumably translated in the cytochrome and kept there during G0-G1. Meanwhile,

Orc2 facilitates binding of the replication licensing factor (RLF) to the chromosomes and they recruit Cdc6, Cdt1 and loading of Mcm2-7 to form origins of replication during G1. In S-phase, replication is initiated at each marked spot and continues until the chromosome is completely replicated exactly one time. When the nuclear membrane is dissolved in prophase the Cdt1/RLF is removed and sequestered by geminin to prevent any further replication [185]. In late telophase, the activity of geminin is disrupted and the possibility of replication licensing returns as the daughter cells form nuclei and enter G0-G1 [186].

It seems very reasonable that a similar process (initiated by the nuclear membrane dissolution giving access to the DNA for a protein left in the cytoplasm of daughter cells in the cytoplasm) sequentially marks GSCKs in the pericentromeric heterochromatin for transcription. The ATRX protein seems to have such a function.[11]

---

[11] McDowell TL et al. 1999. Localization of a putative transcriptional regulator (ATRX) at pericentromeric heterochromatin and the short arms of acrocentric chromosomes. *Proc Natl Acad Sci USA.* 96(24):13983-8.

He Q et al. 2015. The Daxx/Atrx Complex Protects Tandem Repetitive Elements during DNA Hypomethylation by Promoting H3K9 Trimethylation. Cell Stem Cell. 17(3):273-86.

ATRX/DXX directs trimethylation of H3K9 to suppress transcription of heterochromatin.

After being transcribed in a generation of the cell (G0), that GSCK would be permanently suppressed. There are a number of proteins that may play roles in such a process including STELLA [187-189], GNMM[190] and IKAROS [191-194].

Differentiation

In adult humans, the master development program has already run its course in most tissue types.[12] However, the immune system is constantly being renewed (top-to–bottom) from its stem cells and it is here that we can most likely observe the execution of the several branches of development program as stem cells give rise to progenitor cells that differentiate to various mature immune cell types.

IKAROS seems to be an archetypical transcription factor for executing the master development program. It is the main representative of a family of zinc-finger

---

[12] This is why the biochemistry of differentiation is relatively unknown. People are not looking in the right place.

transcription factors [195-203]. It was first recognized as critical to the differentiation of leukocyte and lymphocyte lineages [204-214]. Moreover, dysfunction of IKAROS is linked to failure of the differentiation program resulting in leukemia [208, 215]. Later IKAROS was discovered to bind to the pericentromeric heterochromatin in a cell-cycle dependent way [191, 193, 194, 201, 216-218]. This function is determined by phosphorylation of IKAROS [191, 193, 201, 216, 219-222]. Most recently, it has been shown that IKAROS is associated with stage-specific (e.g., generation-specific), not linage specific, differentiation [191, 205, 212, 223-226].

The behavior of IKAROS is more or less exactly what I would expect in the execution of the MDP [191, 205, 223, 227-229].

If you want to unravel the Master Development Program, this is the correct place to start.[13]

---

[13] Georgopoulos K, Winandy S, Avitahl N. 1997. The role of the Ikaros gene in lymphocyte development and homeostasis. *Annu Rev Immunol.* 15:155-76.

Li Z et al. 2012. Cell cycle-specific function of Ikaros in human leukemia. Pediatr *Blood Cancer.* 59(1):69-76.

Part 2. Differentiation and Asymmetric Division

Complex animals are composed of many clones of different cell types. It is clear that a single cell (i.e., the zygote) gives rise to all the different cell types in complex animals. How does this happen?

The Cell Cycle

One of the first successes of biology was to explain cell division (mitosis). The principal stages and phases of the cell cycle are described as follows:

> G0 - The cell is active in doing what the cell was designed to do, i.e., transcribing mRNA from protein-coding genes and conducting basal metabolism. During this stage, the euchromatic DNA is deployed for transcription and is subject to frequent single-strand breaks and double-strand breaks. When external replication signals are received or when DNA damage reaches an excessive level, the cell enters the repair and replication cycle. The overall stress on a cell is determined by metabolic, mechanical and genomic factors, which are sensed by separate systems each of which activates p53. The level of p53 activation appears to

determine the immediate fate of that particular cell: replication, apoptosis or senescence.

G1 - The cell begins re-condensing and repairing its chromosomes. To the extent possible, damage is identified and repaired by various non-homogeneous mechanisms. As damage is identified (e.g., by the ATM protein). The extent of damage is registered as "genomic stress" by activation of the p53 protein, which initially activates p21 protein. Protein p21 slows the rate of retrieval (e.g., rewinding) of the DNA to give more time for DNA repair mechanisms to work. If this is unsuccessful in mitigating the stress, higher levels of p53 activation activate Bax and other pro-apoptotic proteins. These proteins are tied up by anti-apoptotic proteins of the same family such as Bcl-2. When this happens, p27 is released to arrest the cell cycle and perhaps route the cell into senescence. (Senescence is particularly likely in the case of depleted telomeres (i.e., terminal cells after numerous cycles of proliferation) or metabolic stress, e.g., not enough oxygen or nutrients). If this does not work, the activation of pro-apoptotic proteins will normally initiate the caspase cascade of proteases, which begin the disassembly and recycling of the cell. Caspase 3 actually attacks Bcl-2 and cleaves it into a pro-apoptotic protein.

If the damage-repair process is successful in producing *normal-looking* chromosomes (which may have various deletions, insertions and/or rearrangements), the chromosomes are readied for replication.

S—The synthesis stage of the cell cycle begins by matching each chromosome with its sister. Then the four strands of the two chromosomes are compared to one another in such a way that the shorter strand (containing deletions) are repaired by using its complement as a template (i.e., homogeneous DNA repair).[14] Once homogeneous repair has produced chromosomes that no longer contain identifiable deletions[15], replication proceeds

---

[14] The odds of all four strands having deletions in the same place are fairly low so this process works nicely; but point mutations in the template strand are incorporated in the repair patch and entire segments of chromosomes are swapped back and forth.

[15] I assume that the pairing of the shortened strand with the corresponding segments of the longer strand creates a loop in the longer strand corresponding to the deleted segment. Spontaneous (non-enzymatic) rupture of the short strand at the point of the deletion allows DNA repair of the gap using the longer strand as the template.

using the existing strands as templates for entirely new strands. In actuality, it is difficult to distinguish homogeneous repair from DNA replication.

As noted above, replication is prevented from going past the duplications stage by involvement of the *replication licensing factor* (i.e., geminin, GMNN), which is presumably translated in the cytochrome and kept there during G0-G1 phases. During G1, Orc2 facilitates binding of the replication licensing factor (RLF) to the chromosomes and they recruit Cdc6, Cdt1 and loading of Mcm2-7 to form origins of replication. In S-phase, replication is initiated at each marked spot and continues until the chromosome is completely replicated exactly one time. When the nuclear membrane is dissolved in prophase the Cdt1/RLF is removed and sequestered by geminin to prevent any further replication [185]. In late telophase, the activity of geminin is disrupted and the possibility of replication licensing returns as the daughter cells form nuclei and enter G0-G1 [186].

G2—The G2 stage is the key to cell identity and differentiation. In simple proliferation, the daughter strands of DNA are marked by DNA methylation to match the parent strands.

M--Appropriate histone marks are accumulated on the nucleosomes and the chromosomes are separated into two identical daughter sells by mitosis.

If the cell makes it through mitosis, but displays a highly abnormal phenotype resulting from genetic or epigenetic errors, it normally attracts the attention of the immune system. Leukocytes may attack the abnormal cell with "death signals" (FAS-L, TNF-alpha, etc.), which activate the caspase cascade (via e.g., the FAS receptor) without involving the internal stress monitoring system.

Simple Differentiation

In the case of simple differentiation, the new G0 stage opens with the transcription of a new GSCK [230-233], which (among other things) modifies the pattern of DNA methylation and, hence, causes differentiation of the transcriptome and proteasome.

It is conceivable that through external signals the clones of differentiating cells could cycle at different rates resulting in mixtures of different types of cells all existing at the same time. But it would be very difficult to control development of such a mass. I suspect that the differentiation observed in tumors may be of this type.

Asymmetric Division and Stem Cells

 In trying to make a system that would not depend in any way on external signals, I was at first baffled by this problem. The literature (circa 2005) was of no help. What I discovered was that the state of the art in asymmetric division mechanisms was *hypothesized* to involve polarization of the cytoplasm (as the result of some external signal or gradient) *followed by* changes in the DNA and this idea is still dominant [234]. Indeed, no one seemed to even think much about how polarization of the cytochrome could fundamentally change the DNA to yield a distinctly different transcriptome. Moreover, it appeared that the polarization that was invoked would cause both of the daughter cells to be different from the parent as well as different from each other. And, of course, what sets up the polarization allegedly preceding all the stem cell divisions?

After contemplating the problem for some time (dedicated to the idea that the key reprogramming must happen in the G2 stage and be guided by nmRNAs) in 2009, I came up with what I think it the most original idea in this hypothesis [9, 12]. I then discovered the immortal strand hypothesis [12] and discovered that my hypothesis was consistent with those observations: Namely, if a stem cell only undergoes asymmetric divisions, the parental strand always is saved intact [235-245]; however, if the stem cell proliferates normally, the parental strand is subject to normal degradation (aging) [246]. Moreover, my system did not require any complex mechanisms as had been proposed by others [247]. The only limitation was that only one chromosome could be modified at a time, which seems very reasonable for fundamental changes in the pattern of differentiation.

Assuming that only one chromosome is modified at a time, the process is outlines on the following pages. When the stem cell is cued to generate a new clone, it enters a cell cycle. After the new strand is synthesized (shown in red in the diagram below), it enters the G2 stage [248, 249]. However, a series of nmRNAs (from a GSCKs [250-254]) have been targeted to its new (blank) strand and guide formation of the new (i.e., different from the parent) epigenetic markings [255-258]. Thus, a cell with strands of DNA with different markings is produced and divides into a first generation of daughter cells with the mis-matched markings. Immediate return to the cell cycle (with normal G2 duplication of markings for each strand) produces double-strands of the target chromosome with two different sets of epigenetic marking.

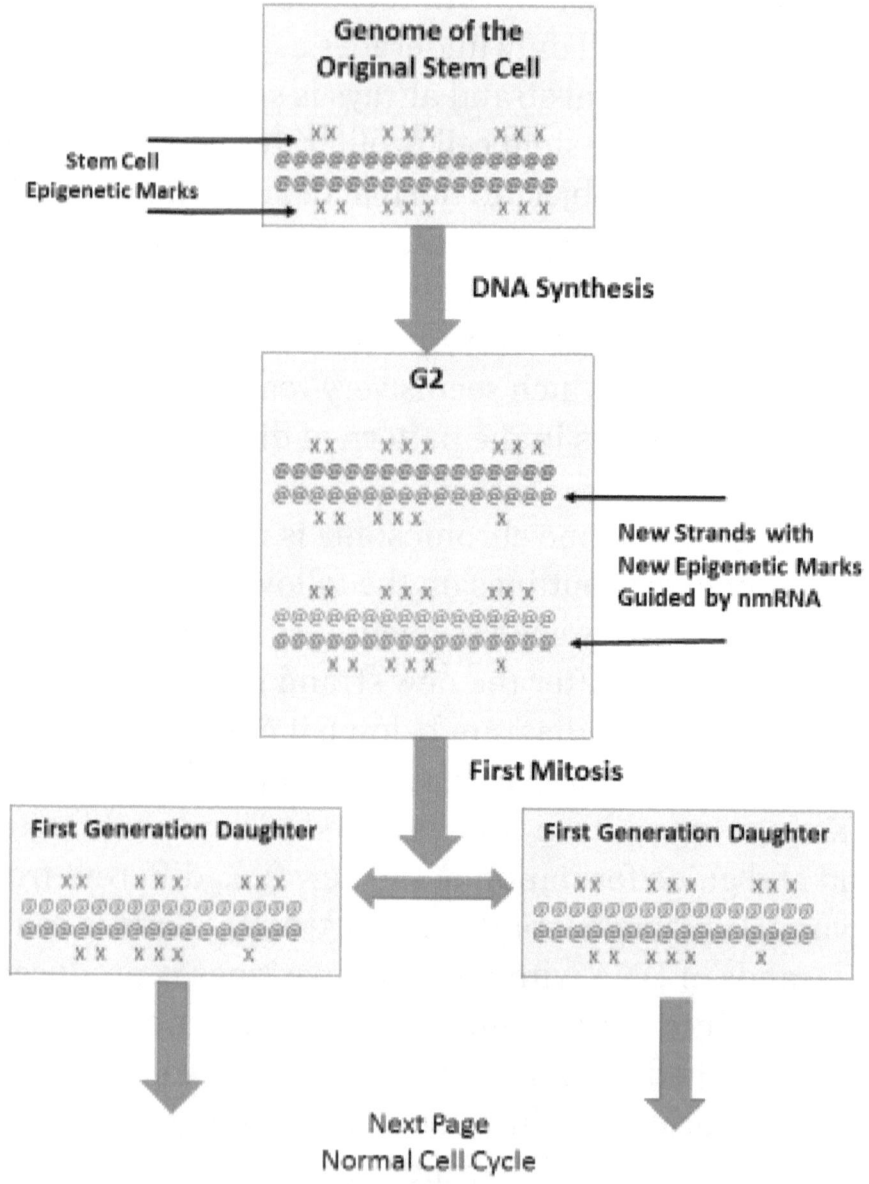

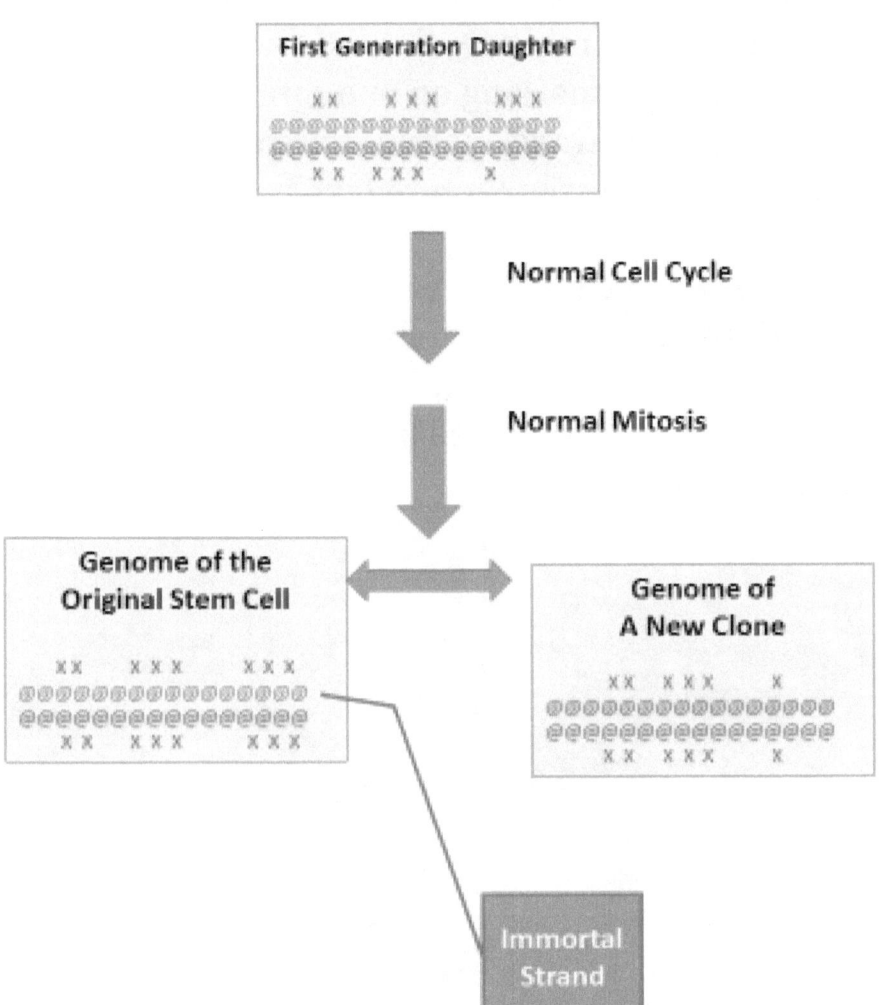

When the second-generation cells form through normal mitosis, the original stem-cell markings regenerate the stem cell and an entirely new clone is founded (with different epigenetic marking on one chromosome). Note that the "immortal strand" *defines the stem cell* and does not have to be selectively captured by it or targeted to it.

I believe that the recent observation concerning histones [259, 260] and labeled bases [243] are entirely consistent with this mechanism. Epigenetic changes in DNA (e.g., methylation) are accompanied by changes in the histones of the nucleosomes [261].

Part 3. Evolution

Evolution and Speciation

In Part 1 (above) I discussed some of the fundamental points I wish to make about evolution and speciation. Contrary to Darwin, I do not see speciation as a *result of evolution*. Speciation and evolution are two separate phenomena. Consistent with Goldschmidt, I think sympatric speciation is the rule rather than the exception [45, 49]. Indeed, even if you have two populations of a species separated geographically and evolving separately, any actual speciation event occurs within a family within one of the populations (i.e., pre-mating barriers to reproduction are unlikely to occur [262]). I have identified pericentric inversions as the most common cause of speciation; although chromosomal fusion or fission, or non-pericentric inversions are also possible events leading to speciation. It only takes one such genomic event and two generations of inbreeding to produce a nascent species [45]. And it is likely that every large population contains many such nascent species (which if separated into a new environment would tend to evolve independently and give the impression of allopatric speciation). Wasserlauft et al. [56] has essentially confirmed this idea with two closely related species of fruit fly (*Drosophila virilis* (*phyla virilis*) and *D. kanekoi* (*phyla montana*)). Cioffi Mde and coworkers have also suspected this as the source of speciation in fish [58]; and, observation of recently separated species (15-20 thousand generations) of whitefish also supports this idea of imbedded nascent species [263].

The Mechanism of Evolution

I did not set out to explain evolution when I tried to develop a hypothesis to explain development, but I was very pleased when I noticed that the mechanism I devised would explain evolution [9, 10] and even the developmental hourglass. In my reasoning, I accepted the idea that "random" mutations are the subject of natural selection; but in higher organisms two factors are needed to focus natural selection on progressive improvement. First, the mutations need to be applied to the beginning or end of the developmental code so that *progress along a course* can be achieved (rather than *meandering* around the starting point). This mechanism is provided by *terminal addition* of codes to the Master Development Program. The second factor that is needed is sexual reproduction, which allows (tentatively successful) mutations to spread rapidly through a population where they can be tested and fixed in the population.

There are a variety of studies that imply *rapid evolution*[16] (changes in phenotype) following sympatric speciation, especially in fish [264, 265] and plants [266]. Interestingly, transposable elements [267-269] (derived from pericentromeric heterochromatin [264, 270, 271]) and microRNAs [272, 273] have been suggested as key players in rapid diversification of species.

Hotspots and Retroviral Pandemics

It is clear that the chromosomes experience mechanical stress as the DNA is moved about by protein fibrils. Assuming that the primary pulling occurs at the centromere, the "viscous friction" (for lack of a better term) resulting from movement of the strand of DNA through the liquid medium will maximize internal stress at the centromere. At the transitions (i) from the *centromere to the pericentromeric heterochromatin* and (ii) from the *pericentromeric heterochromatin to the euchromatin*, the DNA strands are relatively weak (compared to the pericentromeric heterochromatin) and are at high risk for double-strand breaks [274-278]. This fact is demonstrated by the high frequency of pericentric inversions (e.g., about 1% of the human population carries an inversion of at least one chromosome [55, 279]) and

---

[16] Most authors refer to "rapid speciation" (assuming that speciation is caused by evolution) when they refer to phenotypic changes observed in different related species.

the presence of unbalanced chromosomes (e.g. human chromosomes Y, 13, 14, 15, 21 and 22). Thus, these zones are hotspots for breakage, which can be incorrectly repaired by inclusion of stray pieces of DNA.

The sources of pericentromeric (satellite) DNA is most likely the products of retro-transcriptions of RNA. There is always some level of reverse-transcriptase enzyme in the population, but it is believed that this is highly magnified during periods of retroviral pandemics. These viruses (especially when associated with sexual activity) are well positioned to introduce retro-transcripts into the genome. Thus, our genomes and the genomes of other species [280, 281] reflect the history of pandemics occurring at various intervals over the last 500-million years, e.g., some sources indicate humans were exposed to 31 retroviral pandemics over the past 60-million years.

As shown in the figure below (and discussed by Hovarth et al. [40, 72, 85, 86, 282]), the random pieces of DNA are inserted (i.e., seeded) into the breakpoints in the DNA on either end of the satellite DNA. These elements are then swapped (among chromosomes) and selected by natural selection to produce new Generation Specific Control Keys (GSCKs).

As indicated in the discussion of the operation of the Master Development Program, the program is apparently executed from the euchromatin towards the centromere. Thus, adverse variations in the early part of the code would have profound adverse effects on reproductive success and the phenotype of the progeny. On the other hand, variations near the centromere, only affect terminal differentiation of cell clones.

As indicated in the figure (below) there are several sources for RNAs to be converted into GSCKs. There is direct experimental evidence that random RNA transcripts are converted into RNA:DNA hybrids by reverse transcriptase and become inserted into pericentric DNA.[17]   In particular, introns [47, 63, 96, 140, 168, 283-296] and promoter associate RNAs [297-299] (intronRNAs and pmRNAs) are likely candidates for the formation of GSCKs because they already contain the sequence of base pairs needed to target their respective genes. Of course, viral elements themselves are interesting as they may insert themselves near genes as well as in the pericentric heterochromatin [112].

---

[17] Bersani F et al. 2015. Pericentromeric satellite repeat expansions through RNA-derived DNA intermediates in cancer. *Proc Natl Acad Sci USA*. 112(49):15148-53.

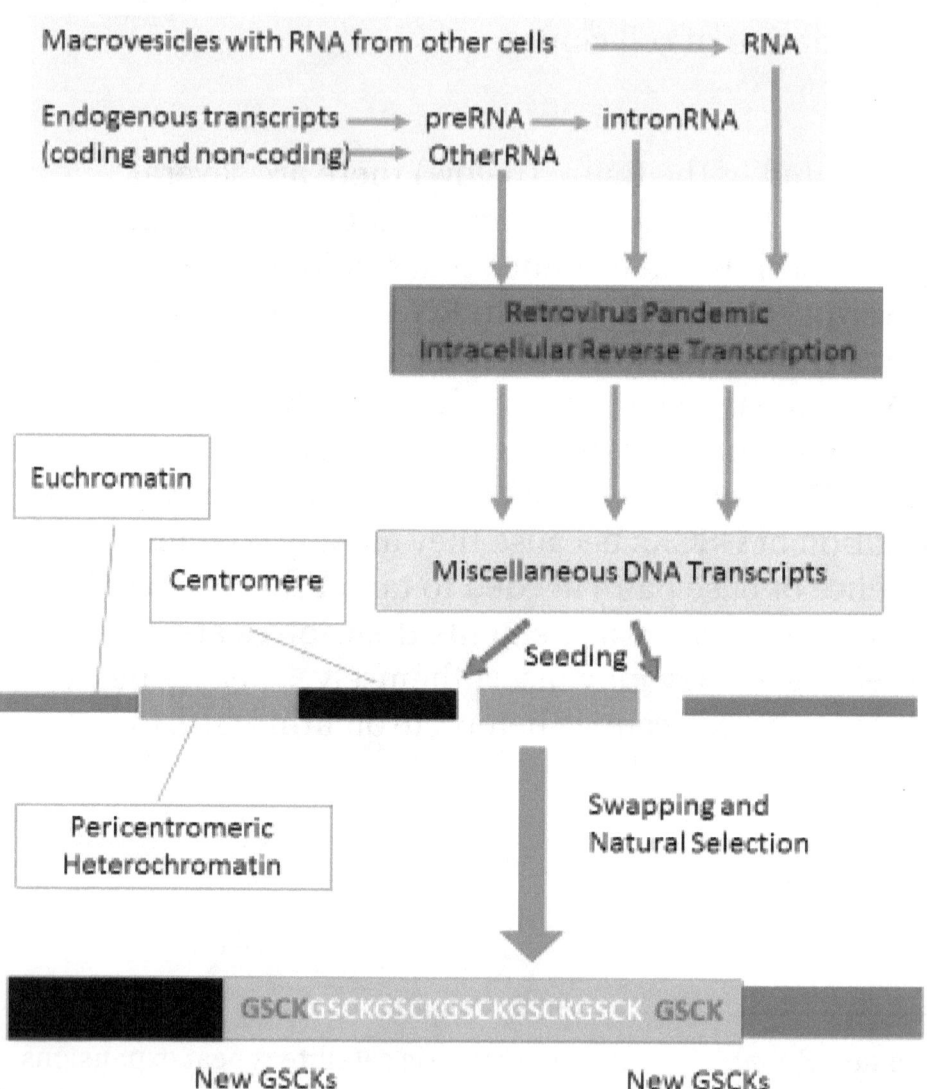

Punctuated Equilibrium

Once again, the MDP hypothesis offers an unexpected answer to an unrelated nagging question: Why is evolution punctuated? It has been noted that stasis in phenotype can persist for very long periods of time and that transitional forms are relatively rare in the fossil record. Hence, evolution appears to happen in "random" burst (i.e., punctuations). This concept was the focus of Stephen J. Gould in the 1980s and 1990s [68, 69] and a variety of efforts have been made to incorporate it into evolutionary theory [300-303]. These factors make it difficult for the Darwinists to fight off the creationists as species seem to appear fully-formed and without the extinction of obviously related species in the fossil record [67, 304]. But, according to the MDP and related hypotheses argued above, the principal species (phenotype) may harbor numerous small populations of nascent species (i.e., the same phenotype, but with homozygous inverted chromosomes).

If you have a primary population with a variety of nascent inbreeding new species (e.g., resulting from homozygous pericentric inversions), a retrovirus pandemic that affects all of a nascent species (not necessarily the entire primary species) facilitates a burst of genomic variation (of phenotype).

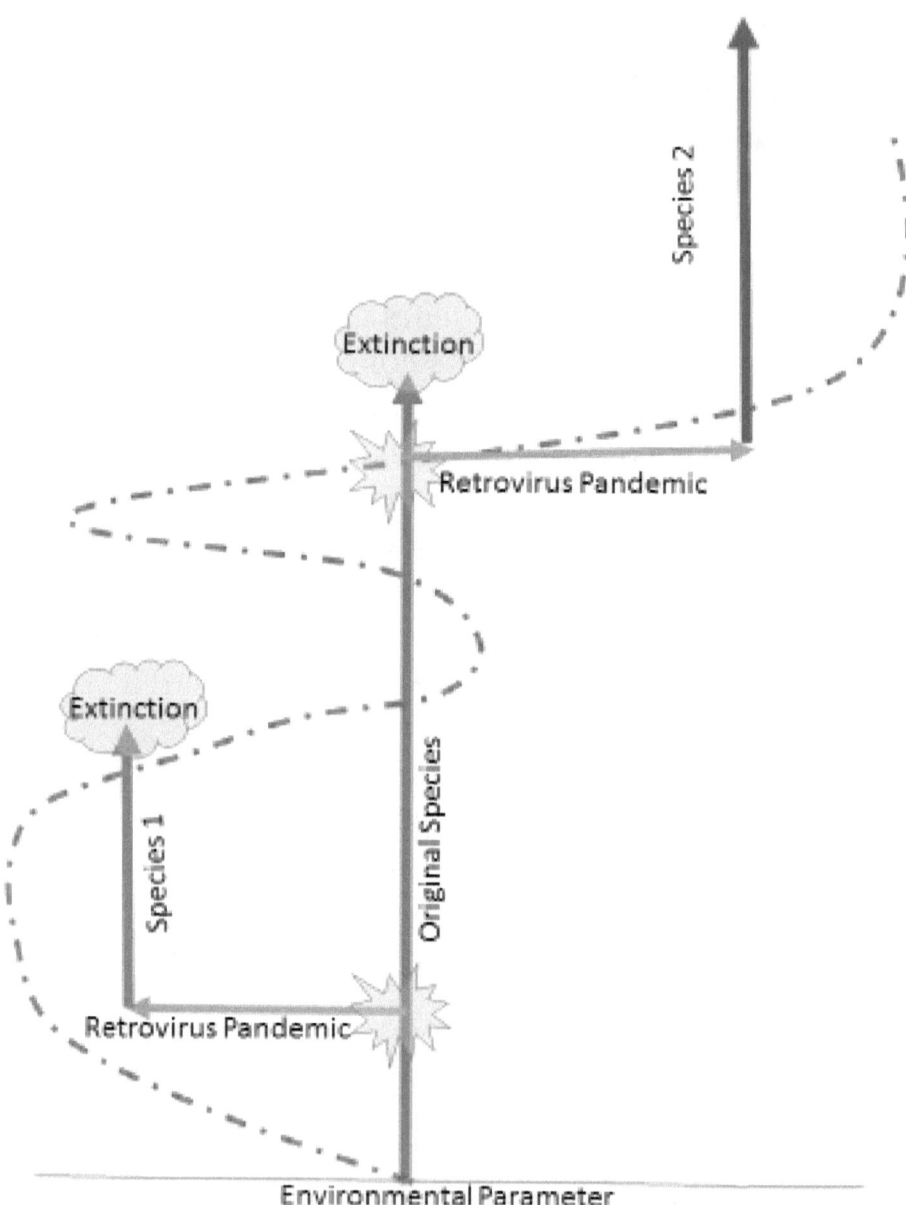

In the much larger primary species, the mutations caused by the retrovirus will soon be lost (through dilution) assuming that the species is adapted to its environment. However, within the much smaller nascent species, the mutations can become fixed.[18]

If environmental variations favor a new phenotype, natural selection will quickly allow the small nascent species to adapt its phenotype to best advantage, while the parent species remains essentially unchanged (in stasis). I have written about this in some detail [11, 45, 49]. The figure above summarizes the process.

---

[18] There is evidence that retrovirus pandemics effect only segments of a population. Human ERV K111 is found in pericentromeric heterochromatin in >97% of Africans, but only about 85% of non-Africans:

Kaplan MH et al. 2019. Structural variation of centromeric endogenous retroviruses in human populations and their impact on cutaneous T-cell lymphoma, Sézary syndrome, and HIV infection. BMC Med Genomics. 12(1):58.

Part 4. Summary Support for the Hypothesis

The piece of information that is most critical to the MDP hypothesis is generally missing. While there is now an abundance of evidence that pericentromeric (satellite) heterochromatin is transcribed [103, 108, 123, 305, 306], there is little *direct* evidence that transcripts from the pericentromeric heterochromatin regulate the genome in each cell cycle (i.e., in each generation of the cell clone).

The best I can point at are the following:

> The epithelial-to-mesenchymal transition requires modification of the pericentromeric heterochromatin [307].

> Transcripts of the pericentromeric heterochromatin have cis- and trans- effects of transcription of coding genes [99, 106, 107, 125, 128, 191, 291, 308-320].

I believe one reason for the lack of data in this area is that in mature adult animals most cell clones are fully (terminally) differentiated. Hence, the master development program has played itself to its end and is no longer functioning. The major exception to this is the blood forming cell lines and immune system where differentiation from high-level stem cells is ongoing even in adult individuals. Thus, it is not surprising that a number of interesting results are associated with the differentiation of blood and immune cells, especially

IKAROS and related zinc-finger transcription factors [191, 193-195, 197-212, 215-217, 219, 220, 222, 223, 225, 228, 229, 287, 321-362].

Nonetheless, there are a number of observations that imply that the pericentromeric heterochromatin is critical to general *development* and hence the *anatomy* of the species (i.e., circumstantial evidence):

> The configuration of the pericentromeric heterochromatin varies in species [28, 33, 37, 38, 56, 264, 363-375].

> The configuration of the pericentromeric heterochromatin varies with development (or reprogramming of stem cells) of a single species [175, 376-382].

> The configuration of pericentromeric heterochromatin varies with differentiation of specific cell clones [175, 264, 383-385].

> The transcription and suppression of transcription of the pericentromeric heterochromatin varies with the cell cycle [101, 191, 250, 386-393].

Most authors seem to associate transcription of the pericentromeric (satellite) heterochromatin with "maintenance of chromosome stability." Obviously, I think that all that transcription of the pericentromeric heterochromatin has a much wider role.

And

> Changes (mutations) in the pericentromeric heterochromatin are frequently associated with various birth defects.[19] Moreover, ERV transcripts are involved with embryonic development.[20]

Overall, I take comfort from the fact that the system does not require external gradients or guidance, it is hard-

---

[19] Malinverni AC et al. 2017. Unusual Duplication in the Pericentromeric Region of Chromosome 9 in a Patient with Phenotypic Alterations. *Cytogenet Genome Res.*150(2):100-105.

Okamoto N et al. 2014. A clinical study of patients with pericentromeric deletion and duplication within 16p12.2-p11.2. Am *J Med Genet A*. 164A(1):213-9.

Déjardin J. 2015. Switching between Epigenetic States at Pericentromeric Heterochromatin. *Trends Genet.* 31(11):661-672.

[20] Deng R et al. 2018. Identification and characterization of ERV transcripts in goat embryos. *Reproduction.* pii: REP-18-0336.R4.

Chen C et al. 2019. Retrotransposons evolution and impact on lncRNA and protein coding genes in pigs. *Mob DNA.* 10:19.

wired and it unifies both development and evolution. Finally, the chemistry is plausible.

Appendix. Larger Scale Copies of Key Graphics

Operation of the
Hypothetical Master Development Program

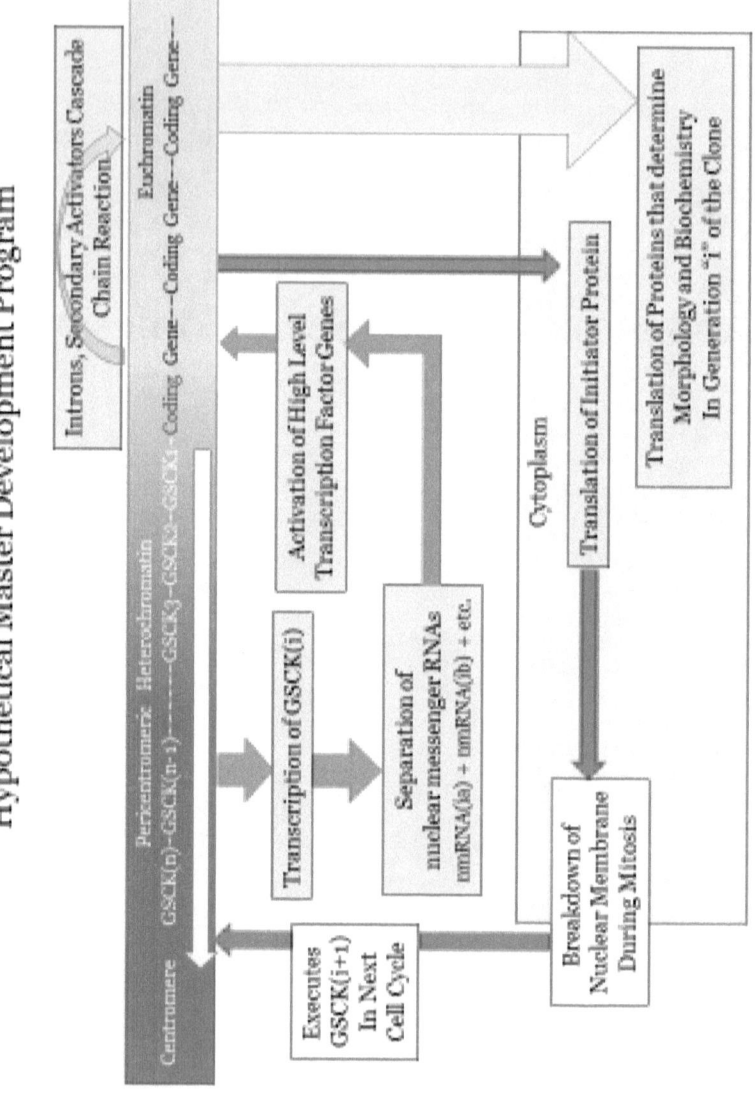

Evolution of the
Hypothetical Master Development Program

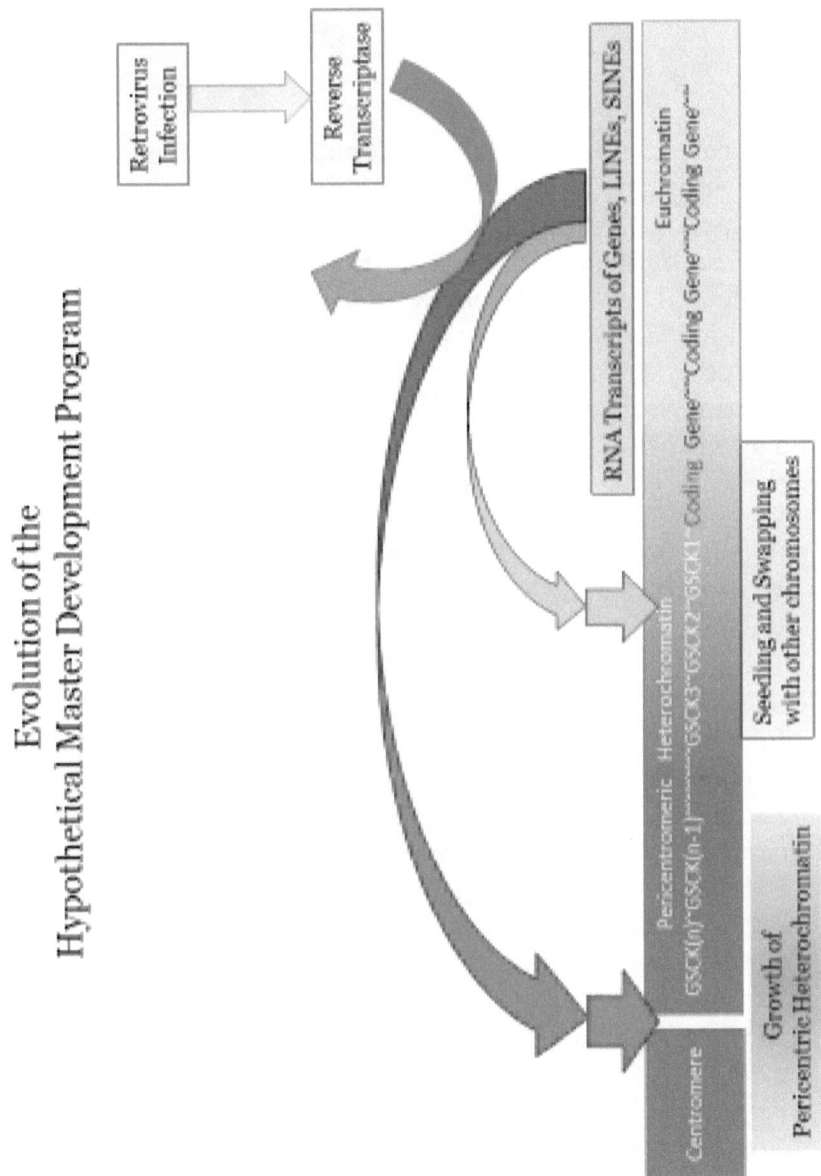

Bibliography

1.   Wolpert, L., *Positional information revisited.* Development, 1989. **107 Suppl**: p. 3-12.
2.   Wolpert, L., *Positional information and the spatial pattern of cellular differentiation.* J Theor Biol, 1969. **25**(1): p. 1-47.
3.   Wolpert, L., *The development of the pattern of growth.* Postgrad Med J, 1978. **54 Suppl 1**: p. 15-24.
4.   Wolpert, L., *Positional information and pattern formation.* Philos Trans R Soc Lond B Biol Sci, 1981. **295**(1078): p. 441-50.
5.   Towers, M., L. Wolpert, and C. Tickle, *Gradients of signalling in the developing limb.* Curr Opin Cell Biol, 2012. **24**(2): p. 181-7.
6.   Wolpert, L., *Diffusible gradients are out - an interview with Lewis Wolpert. Interviewed by Richardson, Michael K.* Int J Dev Biol, 2009. **53**(5-6): p. 659-62.
7.   Wolpert, L., *Positional information and patterning revisited.* J Theor Biol, 2011. **269**(1): p. 359-65.
8.   Sheeba, C.J., R.P. Andrade, and I. Palmeirim, *Limb patterning: from signaling gradients to molecular oscillations.* J Mol Biol, 2014. **426**(4): p. 780-4.
9.   Parris, G.E., *A hypothetical Master Development Program for multi-cellular orgnisms: Ontogeny and phylogeny.* Biosciences Hypotheses, 2009. **2**(1): p. 3-12.
10. Parris, G.E., *Developmental diseases and the hypothetical Master Development Program.* Med Hypotheses, 2010. **74**(3): p. 564-73.
11. Parris, G.E., *Scope of medical implications of the Master Development Program hypothesis.* Med Hypotheses, 2010. **74**(5): p. 953.
12. Parris, G.E., *Asymmetric division, stem cells and the immortal strand hypothesis.* Hypot Life Sci, 2011. **1**(2): p. 52-55.
13. Bergman, G., *Pangenesis as a source of new genetic information. The history of a now disproven theory.* Riv Biol, 2006. **99**(3): p. 425-43.
14. Bulmer, M., *The development of Francis Galton's ideas on the mechanism of heredity.* J Hist Biol, 1999. **32**(2): p. 263-92.

15.   Charlesworth, B. and D. Charlesworth, *Darwin and genetics.* Genetics, 2009. **183**(3): p. 757-66.

16.   Geison, G.L., *Darwin and heredity: the evolution of his hypothesis of pangenesis.* J Hist Med Allied Sci, 1969. **24**(4): p. 375-411.

17.   Lenay, C., *Hugo De Vries: from the theory of intracellular pangenesis to the rediscovery of Mendel.* C R Acad Sci III, 2000. **323**(12): p. 1053-60.

18.   Stamhuis, I.H., *The reactions on Hugo de Vries's Intracellular pangenesis: the discussion with August Weismann.* J Hist Biol, 2003. **36**(1): p. 119-52.

19.   Stamhuis, I.H., O.G. Meijer, and E.J. Zevenhuizen, *Hugo de Vries on heredity, 1889-1903. Statistics, Mendelian laws, pangenes, mutations.* Isis, 1999. **90**(2): p. 238-67.

20.   Sturtevant, A.H., C.B. Bridges, and T.H. Morgan, *The Spatial Relations of Genes.* Proc Natl Acad Sci U S A, 1919. **5**(5): p. 168-73.

21.   Basu, A., et al., *Fetus in fetu or differentiated teratomas?* Indian J Pathol Microbiol, 2006. **49**(4): p. 563-5.

22.   Lachapelle, M.Y. and G. Drouin, *Inactivation dates of the human and guinea pig vitamin C genes.* Genetica, 2011. **139**(2): p. 199-207.

23.   Ho, J.W., et al., *Comparative analysis of metazoan chromatin organization.* Nature, 2014. **512**(7515): p. 449-52.

24.   Pita, M., et al., *A highly conserved pericentromeric domain in human and gorilla chromosomes.* Cytogenet Genome Res, 2009. **126**(3): p. 253-8.

25.   Rudd, M.K., G.A. Wray, and H.F. Willard, *The evolutionary dynamics of alpha-satellite.* Genome Res, 2006. **16**(1): p. 88-96.

26.   Feliciello, I., et al., *Satellite DNA as a driver of population divergence in the red flour beetle Tribolium castaneum.* Genome Biol Evol, 2015. **7**(1): p. 228-39.

27.   Kopecna, O., et al., *Tribe-specific satellite DNA in non-domestic Bovidae.* Chromosome Res, 2014.

28.   Vittorazzi, S., L. Lourenco, and S. Recco-Pimentel, *Long-time evolution and highly dynamic satellite DNA in leptodactylid and hylodid frogs.* BMC Genet, 2014. **15**(1): p. 111.

29.   Baicharoen, S., et al., *Locational diversity of alpha satellite DNA and intergeneric hybridization aspects in the Nomascus and Hylobates genera of small apes.* PLoS One, 2014. **9**(10): p. e109151.

30.   Koga, A., et al., *Evolutionary origin of higher-order repeat structure in alpha-satellite DNA of primate centromeres.* DNA Res, 2014. **21**(4): p. 407-15.

31.   Feliciello, I., et al., *Satellite DNA as a Driver of Population Divergence in the Red Flour Beetle Tribolium castaneum.* Genome Biol Evol, 2014. **7**(1): p. 228-39.

32.   Prakhongcheep, O., et al., *Two Types of Alpha Satellite DNA in Distinct Chromosomal Locations in Azara's Owl Monkey.* DNA Res, 2013. **20**(3): p. 235-40.

33.   Chaiprasertsri, N., et al., *Highly Species-Specific Centromeric Repetitive DNA Sequences in Lizards: Molecular Cytogenetic Characterization of a Novel Family of Satellite DNA Sequences Isolated from the Water Monitor Lizard (Varanus salvator macromaculatus, Platynota).* J Hered, 2013. **104**(6): p. 798-806.

34.   Ferree, P.M. and S. Prasad, *How can satellite DNA divergence cause reproductive isolation? Let us count the chromosomal ways.* Genet Res Int, 2012. **2012**: p. 430136.

35.   Pang, A.W., et al., *Mechanisms of Formation of Structural Variation in a Fully Sequenced Human Genome.* Hum Mutat, 2012.

36.   Hara, T., et al., *Tandem repeat sequences evolutionarily related to SVA-type retrotransposons are expanded in the centromere region of the western hoolock gibbon, a small ape.* J Hum Genet, 2012. **57**(12): p. 760-5.

37.   Mravinac, B. and M. Plohl, *Parallelism in evolution of highly repetitive DNAs in sibling species.* Mol Biol Evol, 2010. **27**(8): p. 1857-67.

38.   Mravinac, B., M. Plohl, and D. Ugarkovic, *Conserved patterns in the evolution of Tribolium satellite DNAs.* Gene, 2004. **332**: p. 169-77.

39. Cardone, M.F., et al., *Evolution of beta satellite DNA sequences: evidence for duplication-mediated repeat amplification and spreading.* Mol Biol Evol, 2004. **21**(9): p. 1792-9.

40. Horvath, J.E., et al., *Using a pericentromeric interspersed repeat to recapitulate the phylogeny and expansion of human centromeric segmental duplications.* Mol Biol Evol, 2003. **20**(9): p. 1463-79.

41. Eichler, E.E., N. Archidiacono, and M. Rocchi, *CAGGG repeats and the pericentromeric duplication of the hominoid genome.* Genome Res, 1999. **9**(11): p. 1048-58.

42. Haaf, T. and H.F. Willard, *Orangutan alpha-satellite monomers are closely related to the human consensus sequence.* Mamm Genome, 1998. **9**(6): p. 440-7.

43. Koo, D.H., et al., *Rapid divergence of repetitive DNAs in Brassica relatives.* Genomics. **97**(3): p. 173-85.

44. Smith, C.H., *Alfred Russel Wallace and the road to natural selection, 1844-1858.* J Hist Biol, 2015. **48**(2): p. 279-300.

45. Parris, G.E., *The hopeful monster finds a mate and founds a new species.* Hypoth Life Sci, 2011. **1**(2): p. 32-37.

46. Morgan, T.H., A.H. Sturtevant, and C.B. Bridge, *The Evidence for the Linear Order of the Genes.* Proc Natl Acad Sci U S A, 1920. **6**(4): p. 162-4.

47. Taft, R.J., M. Pheasant, and J.S. Mattick, *The relationship between non-protein-coding DNA and eukaryotic complexity.* Bioessays, 2007. **29**(3): p. 288-99.

48. Mattick, J.S., *A new paradigm for developmental biology.* J Exp Biol, 2007. **210**(Pt 9): p. 1526-47.

49. Parris, G.E., *Application of a hypothesis to speciation in Hominidae.* Hypot Life Sci, 2013. **3**(1): p. 1.

50. Dietrich, M.R., *Richard Goldschmidt: hopeful monsters and other 'heresies'.* Nat Rev Genet, 2003. **4**(1): p. 68-74.

51. Mayr, E., *Speciation and selection.* Proc Am Philos Soc, 1949. **93**(6): p. 514-9.

52. Mayr, E., *Processes of speciation in animals.* Prog Clin Biol Res, 1982. **96**: p. 1-19.

53.   Dutta, U.R., I. Hansmann, and D. Schlote, *Molecular Cytogenetic Characterization of a familial pericentric inversion 3 associated with short stature.* Eur J Med Genet, 2015.

54.   Balay, L., et al., *A familial pericentric inversion of chromosome 11 associated with a microdeletion of 163kb and microduplication of 288kb at 11p13 and 11q22.3 without aniridia or eye anomalies.* Am J Med Genet A, 2015.

55.   Sipek, A., Jr., et al., *Pericentric Inversion of Human Chromosome 9 Epidemiology Study in Czech Males and Females.* Folia Biol (Praha), 2015. **61**(4): p. 140-146.

56.   Wasserlauf, I., et al., *Specific features in linear and spatial organizations of pericentromeric heterochromatin regions in polytene chromosomes of the closely related species Drosophila virilis and D. kanekoi (Diptera: Drosophilidae).* Genetica, 2015. **143**(3): p. 331-42.

57.   Szamalek, J.M., et al., *Characterization of the human lineage-specific pericentric inversion that distinguishes human chromosome 1 from the homologous chromosomes of the great apes.* Hum Genet, 2006. **120**(1): p. 126-38.

58.   Cioffi Mde, B., et al., *Genomic Organization of Repetitive DNA Elements and Its Implications for the Chromosomal Evolution of Channid Fishes (Actinopterygii, Perciformes).* PLoS One, 2015. **10**(6): p. e0130199.

59.   Khadem, M., R. Camacho, and C. Nobrega, *Studies of the species barrier between Drosophila subobscura and D. madeirensis V: the importance of sex-linked inversion in preserving species identity.* J Evol Biol, 2011. **24**(6): p. 1263-73.

60.   Mattick, J.S., *RNA driving the epigenetic bus.* EMBO J, 2012. **31**(3): p. 515-6.

61.   Barry, G. and J.S. Mattick, *The role of regulatory RNA in cognitive evolution.* Trends Cogn Sci, 2012.

62.   Mattick, J.S., *The central role of RNA in the genetic programming of complex organisms.* An Acad Bras Cienc, 2010. **82**(4): p. 933-9.

63.   Mattick, J.S. and M.J. Gagen, *The evolution of controlled multitasked gene networks: the role of introns and other noncoding*

RNAs in the development of complex organisms. Mol Biol Evol, 2001. **18**(9): p. 1611-30.

64.   Keyte, A.L. and K.K. Smith, *Heterochrony and developmental timing mechanisms: changing ontogenies in evolution.* Semin Cell Dev Biol, 2014. **34**: p. 99-107.

65.   Hall, B.K., *Evo-Devo: evolutionary developmental mechanisms.* Int J Dev Biol, 2003. **47**(7-8): p. 491-5.

66.   Hinchliffe, J.R., *Developmental basis of limb evolution.* Int J Dev Biol, 2002. **46**(7): p. 835-45.

67.   Hunt, G., *Evolution in fossil lineages: paleontology and The Origin of Species.* Am Nat, 2010. **176 Suppl 1**: p. S61-76.

68.   Eldredge, N. and S.J. Gould, *On punctuated equilibria.* Science, 1997. **276**(5311): p. 338-41.

69.   Gould, S.J. and N. Eldredge, *Punctuated equilibrium comes of age.* Nature, 1993. **366**(6452): p. 223-7.

70.   Oliver, K.R. and W.K. Greene, *Transposable elements: powerful facilitators of evolution.* Bioessays, 2009. **31**(7): p. 703-14.

71.   Zeh, D.W., J.A. Zeh, and Y. Ishida, *Transposable elements and an epigenetic basis for punctuated equilibria.* Bioessays, 2009. **31**(7): p. 715-26.

72.   Horvath, J.E., et al., *Punctuated duplication seeding events during the evolution of human chromosome 2p11.* Genome Res, 2005. **15**(7): p. 914-27.

73.   Wohr, G., T. Fink, and G. Assum, *A palindromic structure in the pericentromeric region of various human chromosomes.* Genome Res, 1996. **6**(4): p. 267-79.

74.   Alexandrov, I., et al., *Alpha-satellite DNA of primates: old and new families.* Chromosoma, 2001. **110**(4): p. 253-66.

75.   Shepelev, V.A., et al., *Annotation of suprachromosomal families reveals uncommon types of alpha satellite organization in pericentromeric regions of hg38 human genome assembly.* Genom Data, 2015. **5**: p. 139-146.

76.   Shepelev, V.A., et al., *The evolutionary origin of man can be traced in the layers of defunct ancestral alpha satellites flanking the active centromeres of human chromosomes.* PLoS Genet, 2009. **5**(9): p. e1000641.

77.   Kazakov, A.E., et al., *Interspersed repeats are found predominantly in the "old" alpha satellite families.* Genomics, 2003. **82**(6): p. 619-27.

78.   Fedoseyeva, V.B. and A.A. Alexandrov, *Large-scale periodicity of nucleosome positioning signal in pericentric regions of chromosomes (Drosophila melanogaster).* J Biomol Struct Dyn, 2013.

79.   Schueler, M.G., et al., *Progressive proximal expansion of the primate X chromosome centromere.* Proc Natl Acad Sci U S A, 2005. **102**(30): p. 10563-8.

80.   Schueler, M.G., et al., *Genomic and genetic definition of a functional human centromere.* Science, 2001. **294**(5540): p. 109-15.

81.   Schueler, M.G. and B.A. Sullivan, *Structural and functional dynamics of human centromeric chromatin.* Annu Rev Genomics Hum Genet, 2006. **7**: p. 301-13.

82.   Thomas, J.W., et al., *Pericentromeric duplications in the laboratory mouse.* Genome Res, 2003. **13**(1): p. 55-63.

83.   Guy, J., et al., *Genomic sequence and transcriptional profile of the boundary between pericentromeric satellites and genes on human chromosome arm 10p.* Genome Res, 2003. **13**(2): p. 159-72.

84.   Bailey, J.A., et al., *Human-specific duplication and mosaic transcripts: the recent paralogous structure of chromosome 22.* Am J Hum Genet, 2002. **70**(1): p. 83-100.

85.   Horvath, J.E., et al., *Lessons from the human genome: transitions between euchromatin and heterochromatin.* Hum Mol Genet, 2001. **10**(20): p. 2215-23.

86.   Horvath, J.E., et al., *Molecular structure and evolution of an alpha satellite/non-alpha satellite junction at 16p11.* Hum Mol Genet, 2000. **9**(1): p. 113-23.

87.   She, X., et al., *The structure and evolution of centromeric transition regions within the human genome.* Nature, 2004. **430**(7002): p. 857-64.

88.   Lu, X., et al., *The retrovirus HERVH is a long noncoding RNA required for human embryonic stem cell identity.* Nat Struct Mol Biol, 2014.

89. Kelley, D.R. and J.L. Rinn, *Transposable elements reveal a stem cell specific class of long noncoding RNAs.* Genome Biol, 2012. **13**(11): p. R107.

90. Gagneux, P. and A. Varki, *Genetic differences between humans and great apes.* Mol Phylogenet Evol, 2001. **18**(1): p. 2-13.

91. Britten, R.J., *DNA sequence insertion and evolutionary variation in gene regulation.* Proc Natl Acad Sci U S A, 1996. **93**(18): p. 9374-7.

92. Springer, M.S., E.H. Davidson, and R.J. Britten, *Retroviral-like element in a marine invertebrate.* Proc Natl Acad Sci U S A, 1991. **88**(19): p. 8401-4.

93. Jurka, J., W. Bao, and K.K. Kojima, *Families of transposable elements, population structure and the origin of species.* Biol Direct, 2011. **6**: p. 44.

94. Schlesinger, S. and S.P. Goff, *Retroviral transcriptional regulation and embryonic stem cells: war and peace.* Mol Cell Biol, 2015. **35**(5): p. 770-7.

95. Campbell, M., H.J. Kung, and Y. Izumiya, *Long Non-Coding RNA and Epigenetic Gene Regulation of KSHV.* Viruses, 2014. **6**(11): p. 4165-4177.

96. Hadjiargyrou, M. and N. Delihas, *The Intertwining of Transposable Elements and Non-Coding RNAs.* Int J Mol Sci, 2013. **14**(7): p. 13307-28.

97. Isbel, L. and E. Whitelaw, *Endogenous retroviruses in mammals: an emerging picture of how ERVs modify expression of adjacent genes.* Bioessays, 2012. **34**(9): p. 734-8.

98. Lisch, D. and J.L. Bennetzen, *Transposable element origins of epigenetic gene regulation.* Curr Opin Plant Biol, 2011. **14**(2): p. 156-61.

99. Pezer, Z., et al., *Satellite DNA-mediated effects on genome regulation.* Genome Dyn, 2012. **7**: p. 153-69.

100. Tay, S.K., J. Blythe, and L. Lipovich, *Global discovery of primate-specific genes in the human genome.* Proc Natl Acad Sci U S A, 2009. **106**(29): p. 12019-24.

101. Wang, J., et al., *Inhibition of activated pericentromeric SINE/Alu repeat transcription in senescent human adult stem cells reinstates self-renewal.* Cell Cycle, 2011. **10**(17): p. 3016-30.

102. Zaballos, M.A., W. Cantero, and N. Azpiazu, *The TALE Transcription Factor Homothorax Functions to Assemble Heterochromatin during Drosophila Embryogenesis.* PLoS One, 2015. **10**(3): p. e0120662.

103. Saksouk, N., E. Simboeck, and J. Dejardin, *Constitutive heterochromatin formation and transcription in mammals.* Epigenetics Chromatin, 2015. **8**: p. 3.

104. Matylla-Kulinska, K., et al., *Functional repeat-derived RNAs often originate from retrotransposon-propagated ncRNAs.* Wiley Interdiscip Rev RNA, 2014.

105. Mestrovic, N., et al., *Conserved DNA Motifs, Including the CENP-B Box-like, Are Possible Promoters of Satellite DNA Array Rearrangements in Nematodes.* PLoS One, 2013. **8**(6): p. e67328.

106. Wang, D., et al., *Transposon-derived and satellite-derived repetitive sequences play distinct functional roles in Mammalian intron size expansion.* Evol Bioinform Online, 2012. **8**: p. 301-19.

107. Pezer, Z., et al., *Transcription of Satellite DNAs in Insects.* Prog Mol Subcell Biol, 2011. **51**: p. 161-78.

108. Vourc'h, C. and G. Biamonti, *Transcription of Satellite DNAs in Mammals.* Prog Mol Subcell Biol, 2011. **51**: p. 95-118.

109. Carone, D.M., et al., *A new class of retroviral and satellite encoded small RNAs emanates from mammalian centromeres.* Chromosoma, 2009. **118**(1): p. 113-25.

110. Pezer, Z. and D. Ugarkovic, *RNA Pol II promotes transcription of centromeric satellite DNA in beetles.* PLoS One, 2008. **3**(2): p. e1594.

111. Lu, J. and D.M. Gilbert, *Cell cycle regulated transcription of heterochromatin in mammals vs. fission yeast: functional conservation or coincidence?* Cell Cycle, 2008. **7**(13): p. 1907-10.

112. Taruscio, D. and L. Manuelidis, *Integration site preferences of endogenous retroviruses.* Chromosoma, 1991. **101**(3): p. 141-56.

113. Elinson, R.P. and L. Kezmoh, *Molecular Haeckel.* Dev Dyn, 2010. **239**(7): p. 1905-18.

114.   Hopwood, N., *Pictures of evolution and charges of fraud: Ernst Haeckel's embryological illustrations.* Isis, 2006. **97**(2): p. 260-301.

115.   Ohno, S., *Why ontogeny recapitulates phylogeny.* Electrophoresis, 1995. **16**(9): p. 1782-6.

116.   Casci, T., *Development: Hourglass theory gets molecular approval.* Nat Rev Genet, 2011. **12**(2): p. 76.

117.   Irie, N. and S. Kuratani, *Comparative transcriptome analysis reveals vertebrate phylotypic period during organogenesis.* Nat Commun, 2011. **2**: p. 248.

118.   Kalinka, A.T., et al., *Gene expression divergence recapitulates the developmental hourglass model.* Nature, 2010. **468**(7325): p. 811-4.

119.   Hazkani-Covo, E., D. Wool, and D. Graur, *In search of the vertebrate phylotypic stage: a molecular examination of the developmental hourglass model and von Baer's third law.* J Exp Zool B Mol Dev Evol, 2005. **304**(2): p. 150-8.

120.   Bininda-Emonds, O.R., J.E. Jeffery, and M.K. Richardson, *Inverting the hourglass: quantitative evidence against the phylotypic stage in vertebrate development.* Proc Biol Sci, 2003. **270**(1513): p. 341-6.

121.   Melamed-Bessudo, C. and A.A. Levy, *Deficiency in DNA methylation increases meiotic crossover rates in euchromatic but not in heterochromatic regions in Arabidopsis.* Proc Natl Acad Sci U S A, 2012. **109**(16): p. E981-8.

122.   Vincenten, N., et al., *The kinetochore prevents centromere-proximal crossover recombination during meiosis.* Elife, 2015. **4**.

123.   Enukashvily, N.I. and N.V. Ponomartsev, *Mammalian satellite DNA: a speaking dumb.* Adv Protein Chem Struct Biol, 2013. **90**: p. 31-65.

124.   Pezer, Z. and D. Ugarkovic, *Transcription of pericentromeric heterochromatin in beetles--satellite DNAs as active regulatory elements.* Cytogenet Genome Res, 2009. **124**(3-4): p. 268-76.

125.   Chan, F.L., et al., *Active transcription and essential role of RNA polymerase II at the centromere during mitosis.* Proc Natl Acad Sci U S A, 2012. **109**(6): p. 1979-84.

126. Almouzni, G. and A.V. Probst, *Heterochromatin maintenance and establishment: lessons from the mouse pericentromere.* Nucleus, 2011. **2**(5): p. 332-8.

127. Casanova, M., et al., *Heterochromatin reorganization during early mouse development requires a single-stranded noncoding transcript.* Cell Rep, 2013. **4**(6): p. 1156-67.

128. Probst, A.V., et al., *A strand-specific burst in transcription of pericentric satellites is required for chromocenter formation and early mouse development.* Dev Cell, 2010. **19**(4): p. 625-38.

129. Zhang, J., et al., *S phase-dependent interaction with DNMT1 dictates the role of UHRF1 but not UHRF2 in DNA methylation maintenance.* Cell Res, 2011. **21**(12): p. 1723-39.

130. Guo, D., et al., *RNAa in action: from the exception to the norm.* RNA Biol, 2014. **11**(10): p. 1221-5.

131. Esquela-Kerscher, A., *The lin-4 microRNA: The ultimate micromanager.* Cell Cycle, 2014. **13**(7): p. 1060-1.

132. Youngman, E.M. and J.M. Claycomb, *From early lessons to new frontiers: the worm as a treasure trove of small RNA biology.* Front Genet, 2014. **5**: p. 416.

133. Jiao, A.L. and F.J. Slack, *RNA-mediated gene activation.* Epigenetics, 2014. **9**(1): p. 27-36.

134. Wang, J., et al., *Identification of small activating RNAs that enhance endogenous OCT4 expression in human mesenchymal stem cells.* Stem Cells Dev, 2015. **24**(3): p. 345-53.

135. Wang, J., et al., *Inducing gene expression by targeting promoter sequences using small activating RNAs.* J Biol Methods, 2015. **2**(1).

136. Wehkalampi, K., et al., *Association of the timing of puberty with a chromosome 2 locus.* J Clin Endocrinol Metab, 2008. **93**(12): p. 4833-9.

137. Cousminer, D.L., et al., *Targeted resequencing of the pericentromere of chromosome 2 linked to constitutional delay of growth and puberty.* PLoS One, 2015. **10**(6): p. e0128524.

138. Beckwith, J.R., *Regulation of the lac operon. Recent studies on the regulation of lactose metabolism in Escherichia coli support the operon model.* Science, 1967. **156**(3775): p. 597-604.

139. Lewis, M., *The lac repressor.* C R Biol, 2005. **328**(6): p. 521-48.
140. Mattick, J.S., *Challenging the dogma: the hidden layer of non-protein-coding RNAs in complex organisms.* Bioessays, 2003. **25**(10): p. 930-9.
141. Gromak, N., *Intronic microRNAs: a crossroad in gene regulation.* Biochem Soc Trans, 2012. **40**(4): p. 759-61.
142. Li, Z., et al., *Exon-intron circular RNAs regulate transcription in the nucleus.* Nat Struct Mol Biol, 2015. **22**(3): p. 256-64.
143. Marquitz, A.R., et al., *Host Gene Expression is Regulated by Two Types of Noncoding RNAs Transcribed from the Epstein-Barr Virus BART Region.* J Virol, 2015.
144. Wilusz, J.E., *Repetitive elements regulate circular RNA biogenesis.* Mob Genet Elements, 2015. **5**(3): p. 1-7.
145. Qu, S., et al., *Circular RNA: A new star of noncoding RNAs.* Cancer Lett, 2015. **365**(2): p. 141-8.
146. Ivanov, A., et al., *Analysis of intron sequences reveals hallmarks of circular RNA biogenesis in animals.* Cell Rep, 2015. **10**(2): p. 170-7.
147. Jeck, W.R., et al., *Circular RNAs are abundant, conserved, and associated with ALU repeats.* Rna, 2013. **19**(2): p. 141-57.
148. Chen, X., et al., *Horizontal transfer of microRNAs: molecular mechanisms and clinical applications.* Protein Cell, 2012. **3**(1): p. 28-37.
149. Cocucci, E., G. Racchetti, and J. Meldolesi, *Shedding microvesicles: artefacts no more.* Trends Cell Biol, 2009. **19**(2): p. 43-51.
150. Coy, S. and L. Vasiljeva, *The exosome and heterochromatin : multilevel regulation of gene silencing.* Adv Exp Med Biol, 2011. **702**: p. 105-21.
151. Gezer, U., et al., *Long non-coding RNAs with low expression levels in cells are enriched in secreted exosomes.* Cell Biol Int, 2014. **38**(9): p. 1076-9.
152. Kosaka, N., et al., *Secretory mechanisms and intercellular transfer of microRNAs in living cells.* J Biol Chem, 2010. **285**(23): p. 17442-52.

153.   Lagana, A., et al., *Extracellular circulating viral microRNAs: current knowledge and perspectives.* Front Genet, 2013. **4**: p. 120.

154.   Lee, Y., S. El Andaloussi, and M.J. Wood, *Exosomes and microvesicles: extracellular vesicles for genetic information transfer and gene therapy.* Hum Mol Genet, 2012.

155.   Ludwig, A.K. and B. Giebel, *Exosomes: Small vesicles participating in intercellular communication.* Int J Biochem Cell Biol, 2012. **44**(1): p. 11-5.

156.   Mittelbrunn, M. and F. Sanchez-Madrid, *Intercellular communication: diverse structures for exchange of genetic information.* Nat Rev Mol Cell Biol, 2012. **13**(5): p. 328-35.

157.   Pan, Q., et al., *Hepatic cell-to-cell transmission of small silencing RNA can extend the therapeutic reach of RNA interference (RNAi).* Gut, 2011.

158.   Rayner, K.J. and E.J. Hennessy, *Extracellular Communication via microRNA: Lipid Particles Have a New Message.* J Lipid Res, 2013.

159.   Record, M., *Intercellular communication by exosomes in placenta: A possible role in cell fusion?* Placenta, 2014.

160.   Zlotorynski, E., *Non-coding RNA: Parasite exosomes deliver RNA to hosts.* Nat Rev Mol Cell Biol, 2014. **16**(1): p. 2.

161.   Chu, J.Y., et al., *Dicer function is required in the metanephric mesenchyme for early kidney development.* Am J Physiol Renal Physiol, 2014.

162.   Rybak-Wolf, A., et al., *A variety of dicer substrates in human and C. elegans.* Cell, 2014. **159**(5): p. 1153-67.

163.   Zheng, G.X., et al., *Dicer-microRNA-Myc circuit promotes transcription of hundreds of long noncoding RNAs.* Nat Struct Mol Biol, 2014. **21**(7): p. 585-90.

164.   Muggenhumer, D., et al., *Drosha protein levels are translationally regulated during Xenopus oocyte maturation.* Mol Biol Cell, 2014. **25**(13): p. 2094-104.

165.   Ma, X., et al., *Long non-coding RNAs: a novel endogenous source for the generation of Dicer-like 1-dependent small RNAs in Arabidopsis thaliana.* RNA Biol, 2014. **11**(4): p. 373-90.

166.   Moreno-Moya, J.M., F. Vilella, and C. Simon, *MicroRNA: key gene expression regulators.* Fertil Steril, 2014. **101**(6): p. 1516-23.

167.   Petri, R., et al., *miRNAs in brain development.* Exp Cell Res, 2013.

168.   Meng, Y., et al., *Introns targeted by plant microRNAs: a possible novel mechanism of gene regulation.* Rice (N Y), 2013. **6**(1): p. 8.

169.   Lee, K.J., et al., *Do human transposable element small RNAs serve primarily as genome defenders or genome regulators?* Mob Genet Elements, 2012. **2**(1): p. 19-25.

170.   Jha, A., M. Mehra, and R. Shankar, *The regulatory epicenter of miRNAs.* J Biosci, 2011. **36**(4): p. 621-38.

171.   Taft, R.J., et al., *Small RNAs derived from snoRNAs.* RNA, 2009. **15**(7): p. 1233-40.

172.   Shiekhattar, R., *Dicer finds a new partner in transcriptional gene silencing.* Mol Cell, 2007. **27**(4): p. 519-20.

173.   Harfe, B.D., et al., *The RNaseIII enzyme Dicer is required for morphogenesis but not patterning of the vertebrate limb.* Proc Natl Acad Sci U S A, 2005. **102**(31): p. 10898-903.

174.   Kanellopoulou, C., et al., *Dicer-deficient mouse embryonic stem cells are defective in differentiation and centromeric silencing.* Genes Dev, 2005. **19**(4): p. 489-501.

175.   Cobb, B.S., et al., *T cell lineage choice and differentiation in the absence of the RNase III enzyme Dicer.* J Exp Med, 2005. **201**(9): p. 1367-73.

176.   Lim, A.K. and B.B. Knowles, *Controlling Endogenous Retroviruses and Their Chimeric Transcripts During Natural Reprogramming in the Oocyte.* J Infect Dis, 2015. **212 Suppl 1**: p. S47-51.

177.   Wei, C., et al., *Transcriptome-wide analysis of small RNA expression in early zebrafish development.* RNA, 2012. **18**(5): p. 915-29.

178.   Zhang, C., *Novel functions for small RNA molecules.* Curr Opin Mol Ther, 2009. **11**(6): p. 641-51.

179.   Hu, J., et al., *Promoter-associated small double-stranded RNA interacts with heterogeneous nuclear ribonucleoprotein A2/B1 to induce transcriptional activation.* Biochem J, 2012. **447**(3): p. 407-16.

180.   Wang, X., et al., *Induction of NANOG expression by targeting promoter sequence with small activating RNA antagonizes retinoic acid-induced differentiation.* Biochem J, 2012. **443**(3): p. 821-8.

181.   Blow, J.J. and R.A. Laskey, *A role for the nuclear envelope in controlling DNA replication within the cell cycle.* Nature, 1988. **332**(6164): p. 546-8.

182.   Gillespie, P.J., et al., *Cell Cycle Synchronization in Xenopus Egg Extracts.* Methods Mol Biol, 2016. **1342**: p. 101-47.

183.   Ge, X.Q. and J.J. Blow, *The licensing checkpoint opens up.* Cell Cycle, 2009. **8**(15): p. 2320-2.

184.   DePamphilis, M.L., et al., *Regulating the licensing of DNA replication origins in metazoa.* Curr Opin Cell Biol, 2006. **18**(3): p. 231-9.

185.   Hodgson, B., et al., *Geminin becomes activated as an inhibitor of Cdt1/RLF-B following nuclear import.* Curr Biol, 2002. **12**(8): p. 678-83.

186.   Dimitrova, D.S., et al., *Mammalian nuclei become licensed for DNA replication during late telophase.* J Cell Sci, 2002. **115**(Pt 1): p. 51-9.

187.   Arakawa, T., et al., *Stella controls chromocenter formation through regulation of Daxx expression in 2-cell embryos.* Biochem Biophys Res Commun, 2015.

188.   Liu, Y.J., T. Nakamura, and T. Nakano, *Essential role of DPPA3 for chromatin condensation in mouse oocytogenesis.* Biol Reprod, 2012. **86**(2): p. 40.

189.   Wossidlo, M., et al., *5-Hydroxymethylcytosine in the mammalian zygote is linked with epigenetic reprogramming.* Nat Commun, 2011. **2**: p. 241.

190.   Sylvestre, E.L., et al., *Investigating the potential of genes preferentially expressed in oocyte to induce chromatin remodeling in somatic cells.* Cell Reprogram, 2010. **12**(5): p. 519-28.

191.   Li, Z., et al., *Cell cycle-specific function of Ikaros in human leukemia.* Pediatr Blood Cancer, 2012. **59**(1): p. 69-76.

192.   Helbling Chadwick, L., et al., *The Mi-2/NuRD complex associates with pericentromeric heterochromatin during S phase in rapidly*

proliferating lymphoid cells. Chromosoma, 2009. **118**(4): p. 445-57.

193.   Gurel, Z., et al., *Recruitment of ikaros to pericentromeric heterochromatin is regulated by phosphorylation.* J Biol Chem, 2008. **283**(13): p. 8291-300.

194.   Cobb, B.S., et al., *Targeting of Ikaros to pericentromeric heterochromatin by direct DNA binding.* Genes Dev, 2000. **14**(17): p. 2146-60.

195.   John, L.B., et al., *Pegasus, the 'atypical' Ikaros family member, influences left-right asymmetry and regulates pitx2 expression.* Dev Biol, 2013.

196.   Payne, M.A., et al., *Zinc finger structure-function in Ikaros Marvin A Payne.* World J Biol Chem, 2011. **2**(6): p. 161-166.

197.   Large, E.E. and L.D. Mathies, *hunchback and Ikaros-like zinc finger genes control reproductive system development in Caenorhabditis elegans.* Dev Biol, 2010. **339**(1): p. 51-64.

198.   Cai, Q., et al., *Helios deficiency has minimal impact on T cell development and function.* J Immunol, 2009. **183**(4): p. 2303-11.

199.   Ezzat, S., S. Yu, and S.L. Asa, *The zinc finger Ikaros transcription factor regulates pituitary growth hormone and prolactin gene expression through distinct effects on chromatin accessibility.* Mol Endocrinol, 2005. **19**(4): p. 1004-11.

200.   McCarty, A.S., et al., *Selective dimerization of a C2H2 zinc finger subfamily.* Mol Cell, 2003. **11**(2): p. 459-70.

201.   Dovat, S., et al., *A common mechanism for mitotic inactivation of C2H2 zinc finger DNA-binding domains.* Genes Dev, 2002. **16**(23): p. 2985-90.

202.   Mayer, W.E., et al., *Identification of two Ikaros-like transcription factors in lamprey.* Scand J Immunol, 2002. **55**(2): p. 162-70.

203.   Honma, Y., et al., *Eos: a novel member of the Ikaros gene family expressed predominantly in the developing nervous system.* FEBS Lett, 1999. **447**(1): p. 76-80.

204.   Yokota, T., *Guest editorial: molecular mechanisms of lymphocyte development: recent findings.* Int J Hematol, 2014. **100**(3): p. 218-9.

205.   Yoshida, T. and K. Georgopoulos, *Ikaros fingers on lymphocyte differentiation.* Int J Hematol, 2014. **100**(3): p. 220-9.

206.   Ferreiros Vidal, I., et al., *Genome-wide identification of Ikaros targets elucidates its contribution to mouse B cell lineage specification and pre-B cell differentiation.* Blood, 2013.

207.   Tinsley, K.W., et al., *Ikaros is required to survive positive selection and to maintain clonal diversity during T-cell development in the thymus.* Blood, 2013. **122**(14): p. 2358-68.

208.   Schjerven, H., et al., *Selective regulation of lymphopoiesis and leukemogenesis by individual zinc fingers of Ikaros.* Nat Immunol, 2013. **14**(10): p. 1073-83.

209.   Fu, W., et al., *A multiply redundant genetic switch 'locks in' the transcriptional signature of regulatory T cells.* Nat Immunol, 2012. **13**(10): p. 972-80.

210.   Nutt, S.L. and B.L. Kee, *The transcriptional regulation of B cell lineage commitment.* Immunity, 2007. **26**(6): p. 715-25.

211.   Kathrein, K.L., et al., *Ikaros induces quiescence and T-cell differentiation in a leukemia cell line.* Mol Cell Biol, 2005. **25**(5): p. 1645-54.

212.   Dumortier, A., et al., *Ikaros regulates neutrophil differentiation.* Blood, 2003. **101**(6): p. 2219-26.

213.   Brown, K.E., et al., *Association of transcriptionally silent genes with Ikaros complexes at centromeric heterochromatin.* Cell, 1997. **91**(6): p. 845-54.

214.   Georgopoulos, K., D.D. Moore, and B. Derfler, *Ikaros, an early lymphoid-specific transcription factor and a putative mediator for T cell commitment.* Science, 1992. **258**(5083): p. 808-12.

215.   Winandy, S., P. Wu, and K. Georgopoulos, *A dominant mutation in the Ikaros gene leads to rapid development of leukemia and lymphoma.* Cell, 1995. **83**(2): p. 289-99.

216.   Li, Z., et al., *Cell cycle-specific function of ikaros in human leukemia.* Pediatr Blood Cancer, 2011.

217.   Popescu, M., et al., *Ikaros stability and pericentromeric localization are regulated by protein phosphatase 1.* J Biol Chem, 2009. **284**(20): p. 13869-80.

218. Koipally, J., et al., *Unconventional potentiation of gene expression by Ikaros.* J Biol Chem, 2002. **277**(15): p. 13007-15.

219. Ma, H., et al., *Regulatory phosphorylation of Ikaros by Bruton's tyrosine kinase.* PLoS One, 2013. **8**(8): p. e71302.

220. Uckun, F.M., et al., *Serine phosphorylation by SYK is critical for nuclear localization and transcription factor function of Ikaros.* Proc Natl Acad Sci U S A, 2012. **109**(44): p. 18072-7.

221. Song, C., et al., *Regulation of Ikaros function by casein kinase 2 and protein phosphatase 1.* World J Biol Chem, 2011. **2**(6): p. 126-31.

222. Jantz, D. and J.M. Berg, *Reduction in DNA-binding affinity of Cys2His2 zinc finger proteins by linker phosphorylation.* Proc Natl Acad Sci U S A, 2004. **101**(20): p. 7589-93.

223. Arenzana, T.L., H. Schjerven, and S.T. Smale, *Regulation of gene expression dynamics during developmental transitions by the Ikaros transcription factor.* Genes Dev, 2015. **29**(17): p. 1801-16.

224. Pang, S.H., S. Carotta, and S.L. Nutt, *Transcriptional Control of Pre-B Cell Development and Leukemia Prevention.* Curr Top Microbiol Immunol, 2014.

225. Sellars, M., P. Kastner, and S. Chan, *Ikaros in B cell development and function.* World J Biol Chem, 2011. **2**(6): p. 132-9.

226. Elliott, J., et al., *Ikaros confers early temporal competence to mouse retinal progenitor cells.* Neuron, 2008. **60**(1): p. 26-39.

227. Schwickert, T.A., et al., *Stage-specific control of early B cell development by the transcription factor Ikaros.* Nat Immunol, 2014. **15**(3): p. 283-93.

228. Alsio, J.M., et al., *Ikaros promotes early-born neuronal fates in the cerebral cortex.* Proc Natl Acad Sci U S A, 2013. **110**(8): p. E716-25.

229. Hirono, K., et al., *Identification of hunchback cis-regulatory DNA conferring temporal expression in neuroblasts and neurons.* Gene Expr Patterns, 2012. **12**(1-2): p. 11-7.

230. Papait, R., et al., *The PHD domain of Np95 (mUHRF1) is involved in large-scale reorganization of pericentromeric heterochromatin.* Mol Biol Cell, 2008. **19**(8): p. 3554-63.

231.   Bachman, K.E., M.R. Rountree, and S.B. Baylin, *Dnmt3a and Dnmt3b are transcriptional repressors that exhibit unique localization properties to heterochromatin.* J Biol Chem, 2001. **276**(34): p. 32282-7.

232.   Li, F., et al., *Two novel proteins, dos1 and dos2, interact with rik1 to regulate heterochromatic RNA interference and histone modification.* Curr Biol, 2005. **15**(16): p. 1448-57.

233.   Zhang, R., et al., *HP1 proteins are essential for a dynamic nuclear response that rescues the function of perturbed heterochromatin in primary human cells.* Mol Cell Biol, 2007. **27**(3): p. 949-62.

234.   Rose, L. and P. Gonczy, *Polarity establishment, asymmetric division and segregation of fate determinants in early C. elegans embryos.* WormBook, 2014: p. 1-43.

235.   Winquist, R.J., et al., *Evaluating the immortal strand hypothesis in cancer stem cells: symmetric/self-renewal as the relevant surrogate marker of tumorigenicity.* Biochem Pharmacol, 2014. **91**(2): p. 129-34.

236.   Yadlapalli, S. and Y.M. Yamashita, *DNA asymmetry in stem cells - immortal or mortal?* J Cell Sci, 2013. **126**(Pt 18): p. 4069-76.

237.   Charville, G.W. and T.A. Rando, *The mortal strand hypothesis: non-random chromosome inheritance and the biased segregation of damaged DNA.* Semin Cell Dev Biol, 2013. **24**(8-9): p. 653-60.

238.   Wakeman, J.A., et al., *The immortal strand hypothesis: still non-randomly segregating opinions.* Biomol Concepts, 2012. **3**(3): p. 203-11.

239.   Bussard, K.M., et al., *Immortalized, premalignant epithelial cell populations contain long-lived, label-retaining cells that asymmetrically divide and retain their template DNA.* Breast Cancer Res, 2010. **12**(5): p. R86.

240.   Tajbakhsh, S., *Stem cell identity and template DNA strand segregation.* Curr Opin Cell Biol, 2008. **20**(6): p. 716-22.

241.   Lansdorp, P.M., *Immortal strands? Give me a break.* Cell, 2007. **129**(7): p. 1244-7.

242.   Rando, T.A., *The immortal strand hypothesis: segregation and reconstruction.* Cell, 2007. **129**(7): p. 1239-43.

243. Karpowicz, P., et al., *Support for the immortal strand hypothesis: neural stem cells partition DNA asymmetrically in vitro.* J Cell Biol, 2005. **170**(5): p. 721-32.
244. Tannenbaum, E., J.L. Sherley, and E.I. Shakhnovich, *Evolutionary dynamics of adult stem cells: comparison of random and immortal-strand segregation mechanisms.* Phys Rev E Stat Nonlin Soft Matter Phys, 2005. **71**(4 Pt 1): p. 041914.
245. Merok, J.R., et al., *Cosegregation of chromosomes containing immortal DNA strands in cells that cycle with asymmetric stem cell kinetics.* Cancer Res, 2002. **62**(23): p. 6791-5.
246. Beerman, I. and D.J. Rossi, *Epigenetic regulation of hematopoietic stem cell aging.* Exp Cell Res, 2014.
247. Lew, D.J., D.J. Burke, and A. Dutta, *The immortal strand hypothesis: how could it work?* Cell, 2008. **133**(1): p. 21-3.
248. Halley-Stott, R.P., et al., *Mitosis Gives a Brief Window of Opportunity for a Change in Gene Transcription.* PLoS Biol, 2014. **12**(7): p. e1001914.
249. Kim, W., M. Choi, and J.E. Kim, *The histone methyltransferase Dot1/DOT1L as a critical regulator of the cell cycle.* Cell Cycle, 2014. **13**(5): p. 726-38.
250. Sugimura, K., et al., *Cell cycle-dependent accumulation of histone H3.3 and euchromatic histone modifications in pericentromeric heterochromatin in response to a decrease in DNA methylation levels.* Exp Cell Res, 2010. **316**(17): p. 2731-46.
251. Guasconi, V., et al., *Preferential association of irreversibly silenced E2F-target genes with pericentromeric heterochromatin in differentiated muscle cells.* Epigenetics, 2010. **5**(8): p. 704-9.
252. Akiyama, T., et al., *Transient bursts of Zscan4 expression are accompanied by the rapid derepression of heterochromatin in mouse embryonic stem cells.* DNA Res, 2015.
253. Muramatsu, D., et al., *Pericentric heterochromatin generated by HP1 interaction-defective histone methyltransferase Suv39h1.* J Biol Chem, 2013.
254. Della Ragione, F., et al., *MeCP2 as a genome-wide modulator: the renewal of an old story.* Front Genet, 2012. **3**: p. 181.

255. Barlow, D.P. and M.S. Bartolomei, *Genomic imprinting in mammals.* Cold Spring Harb Perspect Biol, 2014. **6**(2).
256. Zhang, G. and S. Pradhan, *Mammalian epigenetic mechanisms.* IUBMB Life, 2014.
257. Wang, Y., et al., *PiRNAs link epigenetic modifications to reprogramming.* Histol Histopathol, 2014.
258. Guerin, T.M., F. Palladino, and V.J. Robert, *Transgenerational functions of small RNA pathways in controlling gene expression in C. elegans.* Epigenetics, 2014. **9**(1): p. 37-44.
259. Tran, V., L. Feng, and X. Chen, *Asymmetric distribution of histones during Drosophila male germline stem cell asymmetric divisions.* Chromosome Res, 2013. **21**(3): p. 255-69.
260. Tran, V., et al., *Asymmetric division of Drosophila male germline stem cell shows asymmetric histone distribution.* Science, 2012. **338**(6107): p. 679-82.
261. Du, Q., et al., *Methyl-CpG-binding domain proteins: readers of the epigenome.* Epigenomics, 2015: p. 1-23.
262. Grant, B.R. and P.R. Grant, *Lack of premating isolation at the base of a phylogenetic tree.* Am Nat, 2002. **160**(1): p. 1-19.
263. Dion-Cote, A.M., et al., *Reproductive isolation in a nascent species pair is associated with aneuploidy in hybrid offspring.* Proc Biol Sci, 2015. **282**(1802).
264. Symonova, R., et al., *Genome differentiation in a species pair of coregonine fishes: an extremely rapid speciation driven by stress-activated retrotransposons mediating extensive ribosomal DNA multiplications.* BMC Evol Biol, 2013. **13**: p. 42.
265. Genner, M.J., et al., *Geographical ancestry of Lake Malawi's cichlid fish diversity.* Biol Lett, 2015. **11**(6): p. 20150232.
266. Osborne, O.G., et al., *Rapid speciation with gene flow following the formation of Mt. Etna.* Genome Biol Evol, 2013. **5**(9): p. 1704-15.
267. Dion-Cote, A.M., et al., *RNA-seq reveals transcriptomic shock involving transposable elements reactivation in hybrids of young lake whitefish species.* Mol Biol Evol, 2014.

268.   Kapusta, A., et al., *Transposable elements are major contributors to the origin, diversification, and regulation of vertebrate long noncoding RNAs.* PLoS Genet, 2013. **9**(4): p. e1003470.

269.   Blass, E., M. Bell, and S. Boissinot, *Accumulation and rapid decay of non-LTR retrotransposons in the genome of the three-spine stickleback.* Genome Biol Evol, 2012. **4**(5): p. 687-702.

270.   Zhao, M. and J. Ma, *Co-evolution of plant LTR-retrotransposons and their host genomes.* Protein Cell, 2013. **4**(7): p. 493-501.

271.   Du, J., et al., *Evolutionary conservation, diversity and specificity of LTR-retrotransposons in flowering plants: insights from genome-wide analysis and multi-specific comparison.* Plant J, 2010. **63**(4): p. 584-98.

272.   Loh, Y.H., S.V. Yi, and J.T. Streelman, *Evolution of microRNAs and the diversification of species.* Genome Biol Evol, 2011. **3**: p. 55-65.

273.   Smith, L.M., et al., *Rapid divergence and high diversity of miRNAs and miRNA targets in the Camelineae.* Plant J, 2015. **81**(4): p. 597-610.

274.   Tchurikov, N.A., et al., *Hot spots of DNA double-strand breaks and genomic contacts of human rDNA units are involved in epigenetic regulation.* J Mol Cell Biol, 2015. **7**(4): p. 366-82.

275.   Webber, C. and C.P. Ponting, *Hotspots of mutation and breakage in dog and human chromosomes.* Genome Res, 2005. **15**(12): p. 1787-97.

276.   Lopez, V., et al., *Cytokinesis breaks dicentric chromosomes preferentially at pericentromeric regions and telomere fusions.* Genes Dev, 2015. **29**(3): p. 322-36.

277.   Vukovic, B., et al., *Correlating breakage-fusion-bridge events with the overall chromosomal instability and in vitro karyotype evolution in prostate cancer.* Cytogenet Genome Res, 2007. **116**(1-2): p. 1-11.

278.   Berardinelli, F., et al., *mBAND and mFISH analysis of chromosomal aberrations and breakpoint distribution in chromosome 1 of AG01522 human fibroblasts that were exposed to radiation of different qualities.* Mutat Res Genet Toxicol Environ Mutagen, 2015. **793**: p. 55-63.

279.   Demirhan, O., et al., *Correlation of clinical phenotype with a pericentric inversion of chromosome 9 and genetic counseling.* Saudi Med J, 2008. **29**(7): p. 946-51.

280.   Komissarov, A.S., et al., *Tandemly repeated DNA families in the mouse genome.* BMC Genomics, 2011. **12**(1): p. 531.

281.   Chalopin, D., et al., *Evolutionary active transposable elements in the genome of the coelacanth.* J Exp Zool B Mol Dev Evol, 2013.

282.   Perelman, P., et al., *A molecular phylogeny of living primates.* PLoS Genet, 2011. **7**(3): p. e1001342.

283.   Deng, J.H., et al., *Gene Silencing In Vitro and In Vivo Using Intronic MicroRNAs.* Methods Mol Biol, 2015. **1218**: p. 321-40.

284.   Gallegos, J.E. and A.B. Rose, *The enduring mystery of intron-mediated enhancement.* Plant Sci, 2015. **237**: p. 8-15.

285.   Heyn, P., et al., *Introns and gene expression: cellular constraints, transcriptional regulation, and evolutionary consequences.* Bioessays, 2015. **37**(2): p. 148-54.

286.   Artieri, C.G. and H.B. Fraser, *Transcript length mediates developmental timing of gene expression across Drosophila.* Mol Biol Evol, 2014. **31**(11): p. 2879-89.

287.   Yoshida, T., et al., *Transcriptional regulation of the Ikzf1 locus.* Blood, 2013. **122**(18): p. 3149-59.

288.   Valadkhan, S. and L.S. Gunawardane, *Role of small nuclear RNAs in eukaryotic gene expression.* Essays Biochem, 2013. **54**: p. 79-90.

289.   Li, H., D. Chen, and J. Zhang, *Analysis of intron sequence features associated with transcriptional regulation in human genes.* PLoS One, 2012. **7**(10): p. e46784.

290.   Frankel, N., *Multiple layers of complexity in cis-regulatory regions of developmental genes.* Dev Dyn, 2012. **241**(12): p. 1857-66.

291.   Brajkovic, J., et al., *Satellite DNA-like elements associated with genes within euchromatin of the beetle Tribolium castaneum.* G3 (Bethesda), 2012. **2**(8): p. 931-41.

292.   Hung, T. and H.Y. Chang, *Long noncoding RNA in genome regulation: prospects and mechanisms.* RNA Biol, 2010. **7**(5): p. 582-5.

293. Ying, S.Y. and S.L. Lin, *Intron-mediated RNA interference and microRNA biogenesis.* Methods Mol Biol, 2009. **487**: p. 387-413.

294. Fablet, M., et al., *Evolutionary origin and functions of retrogene introns.* Mol Biol Evol, 2009. **26**(9): p. 2147-56.

295. Louro, R., A.S. Smirnova, and S. Verjovski-Almeida, *Long intronic noncoding RNA transcription: expression noise or expression choice?* Genomics, 2009. **93**(4): p. 291-8.

296. Fedorova, L. and A. Fedorov, *Introns in gene evolution.* Genetica, 2003. **118**(2-3): p. 123-31.

297. Taft, R.J., et al., *The relationship between transcription initiation RNAs and CCCTC-binding factor (CTCF) localization.* Epigenetics Chromatin, 2011. **4**: p. 13.

298. Taft, R.J., et al., *Evolution, biogenesis and function of promoter-associated RNAs.* Cell Cycle, 2009. **8**(15): p. 2332-8.

299. Taft, R.J., et al., *Tiny RNAs associated with transcription start sites in animals.* Nat Genet, 2009. **41**(5): p. 572-8.

300. Wallace, R., *Expanding the modern synthesis.* C R Biol, 2010. **333**(10): p. 701-9.

301. Lewontin, R., *The triumph of Stephen Jay Gould.* New York Rev Books, 2008. **55**(2): p. 39-41.

302. Wilkins, J.S. and G.J. Nelson, *Tremaux on species: a theory of allopatric speciation (and punctuated equilibrium) before Wagner.* Hist Philos Life Sci, 2008. **30**(2): p. 179-205.

303. Kutschera, U. and K.J. Niklas, *The modern theory of biological evolution: an expanded synthesis.* Naturwissenschaften, 2004. **91**(6): p. 255-76.

304. Girard, C. and S. Renaud, *Disparity changes in 370 Ma Devonian fossils: the signature of ecological dynamics?* PLoS One, 2012. **7**(4): p. e36230.

305. Chan, F.L. and L.H. Wong, *Transcription in the maintenance of centromere chromatin identity.* Nucleic Acids Res, 2012. **40**(22): p. 11178-88.

306. Mudge, J.M. and M.S. Jackson, *Evolutionary implications of pericentromeric gene expression in humans.* Cytogenet Genome Res, 2005. **108**(1-3): p. 47-57.

307.   Millanes-Romero, A., et al., *Regulation of heterochromatin transcription by Snail1/LOXL2 during epithelial-to-mesenchymal transition*. Mol Cell, 2013. **52**(5): p. 746-57.

308.   Morozov, V.M., et al., *Dualistic function of Daxx at centromeric and pericentromeric heterochromatin in normal and stress conditions*. Nucleus, 2012. **3**(3): p. 276-85.

309.   Collins, A., et al., *RUNX transcription factor-mediated association of Cd4 and Cd8 enables coordinate gene regulation*. Immunity, 2011. **34**(3): p. 303-14.

310.   Lejeune, E., E.H. Bayne, and R.C. Allshire, *On the connection between RNAi and heterochromatin at centromeres*. Cold Spring Harb Symp Quant Biol, 2010. **75**: p. 275-83.

311.   Eymery, A., M. Callanan, and C. Vourc'h, *The secret message of heterochromatin: new insights into the mechanisms and function of centromeric and pericentric repeat sequence transcription*. Int J Dev Biol, 2009. **53**(2-3): p. 259-68.

312.   Ferreira, D., et al., *Satellite non-coding RNAs: the emerging players in cells, cellular pathways and cancer*. Chromosome Res, 2015.

313.   Sjogren, R.J., et al., *Temporal analysis of reciprocal miRNA-mRNA expression patterns predicts regulatory networks during differentiation in human skeletal muscle cells*. Physiol Genomics, 2014: p. physiolgenomics.00037.2014.

314.   Sousa-Victor, P., P. Munoz-Canoves, and E. Perdiguero, *Regulation of skeletal muscle stem cells through epigenetic mechanisms*. Toxicol Mech Methods, 2011. **21**(4): p. 334-42.

315.   Navarro, P., et al., *Molecular coupling of Tsix regulation and pluripotency*. Nature, 2010. **468**(7322): p. 457-60.

316.   Pezer, Z., et al., *Satellite DNA-mediated effects on genome regulation*. Genome Dyn, 2010. **7**: p. 153-69.

317.   O'Neill, R.J. and D.M. Carone, *The role of ncRNA in centromeres: a lesson from marsupials*. Prog Mol Subcell Biol, 2009. **48**: p. 77-101.

318.   Tomilin, N.V., *Regulation of mammalian gene expression by retroelements and non-coding tandem repeats*. Bioessays, 2008. **30**(4): p. 338-48.

319.   Liu, X., et al., *Functional sequestration of transcription factor activity by repetitive DNA*. J Biol Chem, 2007. **282**(29): p. 20868-76.
320.   Ugarkovic, D., *Functional elements residing within satellite DNAs*. EMBO Rep, 2005. **6**(11): p. 1035-9.
321.   Bottardi, S., L. Mavoungou, and E. Milot, *IKAROS: a multifunctional regulator of the polymerase II transcription cycle*. Trends Genet, 2015. **31**(9): p. 500-8.
322.   Nagpal, K., et al., *Ikaros represses protein phosphatase 2A (PP2A) expression through an intronic binding site*. J Biol Chem, 2014.
323.   Boller, S. and R. Grosschedl, *The regulatory network of B-cell differentiation: a focused view of early B-cell factor 1 function*. Immunol Rev, 2014. **261**(1): p. 102-15.
324.   Bottardi, S., et al., *The IKAROS interaction with a complex including chromatin remodeling and transcription elongation activities is required for hematopoiesis*. PLoS Genet, 2014. **10**(12): p. e1004827.
325.   Wang, H., et al., *Pathogenesis and regulation of cellular proliferation in acute lymphoblastic leukemia - the role of Ikaros*. J buon, 2014. **19**(1): p. 22-8.
326.   Tinsley, K.W., et al., *Ikaros is required to survive positive selection and to maintain clonal diversity during T cell development in the thymus*. Blood, 2013.
327.   Rieder, S.A. and E.M. Shevach, *Eos, goddess of treg cell reprogramming*. Immunity, 2013. **38**(5): p. 849-50.
328.   Papatriantafyllou, M., *Haematopoiesis: Two versions of the Ikaros tale*. Nat Rev Immunol, 2013. **13**(11): p. 772-3.
329.   Dovat, S., *Ikaros: the enhancer makes the difference*. Blood, 2013. **122**(18): p. 3091-2.
330.   Wong, L.Y., J.K. Hatfield, and M.A. Brown, *Ikaros sets the potential for Th17-lineage gene expression through effects on chromatin state in early T cell development*. J Biol Chem, 2013.
331.   Heizmann, B., P. Kastner, and S. Chan, *Ikaros is absolutely required for pre-B cell differentiation by attenuating IL-7 signals*. J Exp Med, 2013.

332.   Wong, L.Y., J.K. Hatfield, and M.A. Brown, *Ikaros Sets the Potential for Th17 Lineage Gene Expression through Effects on Chromatin State in Early T Cell Development.* J Biol Chem, 2013. **288**(49): p. 35170-9.

333.   Taylor, N. and V.S. Zimmermann, *Ikaros' suspended flight in the thymus.* Blood, 2013. **122**(14): p. 2291-2.

334.   Rao, K.N., et al., *Ikaros limits basophil development by suppressing C/EBP-alpha expression.* Blood, 2013. **122**(15): p. 2572-81.

335.   Errico, A., *Haematological cancer: Ikaros-not a myth for myeloma.* Nat Rev Clin Oncol, 2013.

336.   Orozco, C.A., et al., *The combined expression patterns of Ikaros isoforms characterize different hematological tumor subtypes.* PLoS One, 2013. **8**(12): p. e82411.

337.   Procko, C., Y. Lu, and S. Shaham, *Sensory Organ Remodeling in Caenorhabditis elegans Requires the Zinc-finger Protein ZTF-16.* Genetics, 2012.

338.   Arranz, L., et al., *Bmi1 is critical to prevent Ikaros-mediated lymphoid priming in hematopoietic stem cells.* Cell Cycle, 2012. **11**(1): p. 65-78.

339.   Oestreich, K.J. and A.S. Weinmann, *Ikaros changes the face of NuRD remodeling.* Nat Immunol, 2012. **13**(1): p. 16-8.

340.   Li, Z., et al., *Ikaros isoforms: The saga continues.* World J Biol Chem, 2011. **2**(6): p. 140-5.

341.   Akimova, T., et al., *Helios expression is a marker of T cell activation and proliferation.* PLoS One, 2011. **6**(8): p. e24226.

342.   Dovat, S., *Ikaros in hematopoiesis and leukemia.* World J Biol Chem, 2011. **2**(6): p. 105-7.

343.   Alinikula, J., et al., *Concerted action of Helios and Ikaros controls the expression of the inositol 5-phosphatase SHIP.* Eur J Immunol, 2010. **40**(9): p. 2599-607.

344.   Ma, S., et al., *Ikaros and Aiolos inhibit pre-B-cell proliferation by directly suppressing c-Myc expression.* Mol Cell Biol, 2010. **30**(17): p. 4149-58.

345.   Matulic, M., et al., *Analysis of Ikaros family splicing variants in human hematopoietic lineages.* Coll Antropol, 2010. **34**(1): p. 59-62.

346.   Komiyama, H., et al., *Alu-derived cis-element regulates tumorigenesis-dependent gastric expression of GASDERMIN B (GSDMB).* Genes Genet Syst, 2010. **85**(1): p. 75-83.

347.   Yoshida, T., S.Y. Ng, and K. Georgopoulos, *Awakening lineage potential by Ikaros-mediated transcriptional priming.* Curr Opin Immunol, 2010. **22**(2): p. 154-60.

348.   Matulic, M., et al., *Ikaros family transcription factors in chronic and acute leukemia.* Am J Hematol, 2009. **84**(6): p. 375-7.

349.   John, L.B., S. Yoong, and A.C. Ward, *Evolution of the Ikaros gene family: implications for the origins of adaptive immunity.* J Immunol, 2009. **182**(8): p. 4792-9.

350.   Meleshko, A.N., et al., *Relative expression of different Ikaros isoforms in childhood acute leukemia.* Blood Cells Mol Dis, 2008. **41**(3): p. 278-83.

351.   Ruiz, A., O. Williams, and H.J. Brady, *The Ikaros splice isoform, Ikaros 6, immortalizes murine haematopoietic progenitor cells.* Int J Cancer, 2008. **123**(6): p. 1240-5.

352.   Ezzat, S. and S.L. Asa, *The emerging role of the Ikaros stem cell factor in the neuroendocrine system.* J Mol Endocrinol, 2008. **41**(2): p. 45-51.

353.   Kiehl, T.R., et al., *Mice lacking the transcription factor Ikaros display behavioral alterations of an anti-depressive phenotype.* Exp Neurol, 2008. **211**(1): p. 107-14.

354.   Agoston, D.V., et al., *Ikaros is expressed in developing striatal neurons and involved in enkephalinergic differentiation.* J Neurochem, 2007. **102**(6): p. 1805-16.

355.   Ghadiri, A., et al., *Critical function of Ikaros in controlling Aiolos gene expression.* FEBS Lett, 2007. **581**(8): p. 1605-16.

356.   Ng, S.Y., T. Yoshida, and K. Georgopoulos, *Ikaros and chromatin regulation in early hematopoiesis.* Curr Opin Immunol, 2007. **19**(2): p. 116-22.

357.   Zhu, X., S.L. Asa, and S. Ezzat, *Ikaros is regulated through multiple histone modifications and deoxyribonucleic acid*

methylation in the pituitary. Mol Endocrinol, 2007. **21**(5): p. 1205-15.

358.   Ezzat, S., et al., *An essential role for the hematopoietic transcription factor Ikaros in hypothalamic-pituitary-mediated somatic growth.* Proc Natl Acad Sci U S A, 2006. **103**(7): p. 2214-9.

359.   Ezzat, S., et al., *Ikaros integrates endocrine and immune system development.* J Clin Invest, 2005. **115**(4): p. 1021-9.

360.   Ling, K.W. and E. Dzierzak, *Ontogeny and genetics of the hemato/lymphopoietic system.* Curr Opin Immunol, 2002. **14**(2): p. 186-91.

361.   O'Neill, D.W., et al., *An ikaros-containing chromatin-remodeling complex in adult-type erythroid cells.* Mol Cell Biol, 2000. **20**(20): p. 7572-82.

362.   Koipally, J., et al., *Repression by Ikaros and Aiolos is mediated through histone deacetylase complexes.* EMBO J, 1999. **18**(11): p. 3090-100.

363.   Bilinski, P., et al., *Diversity and evolution of centromere repeats in the maize genome.* Chromosoma, 2015. **124**(1): p. 57-65.

364.   Sanseverino, W., et al., *Transposon insertion, structural variations and SNPs contribute to the evolution of the melon genome.* Mol Biol Evol, 2015.

365.   Hill, T., et al., *Ultra-High Density, Transcript-Based Genetic Maps of Pepper Define Recombination in the Genome and Synteny Among Related Species.* G3 (Bethesda), 2015.

366.   Zhang, Y., et al., *Chromosomal structures and repetitive sequences divergence in Cucumis species revealed by comparative cytogenetic mapping.* BMC Genomics, 2015. **16**(1): p. 730.

367.   Nergadze, S.G., et al., *Discovery and Comparative Analysis of a Novel Satellite, EC137, in Horses and Other Equids.* Cytogenet Genome Res, 2014.

368.   Rovatsos, M.T., et al., *Molecular and Physical Characterization of the Complex Pericentromeric Heterochromatin of the Vole Species Microtus thomasi.* Cytogenet Genome Res, 2014.

369.   Nwakanma, D.C., et al., *Breakdown in the Process of Incipient Speciation in Anopheles gambiae.* Genetics, 2013.

370. Bueno, D., O.M. Palacios-Gimenez, and D.C. Cabral-de-Mello, *Chromosomal Mapping of Repetitive DNAs in the Grasshopper Reveal Possible Ancestry of the B Chromosome and H3 Histone Spreading.* PLoS One, 2013. **8**(6): p. e66532.

371. de Franca Rocha, M., N.F. de Melo, and M.J. de Souza, *Comparative cytogenetic analysis of two grasshopper species of the tribe Abracrini (Ommatolampinae, Acrididae).* Genet Mol Biol, 2011. **34**(2): p. 214-9.

372. Rovatsos, M.T., et al., *Rapid, independent, and extensive amplification of telomeric repeats in pericentromeric regions in karyotypes of arvicoline rodents.* Chromosome Res, 2011.

373. Devos, K.M., *Grass genome organization and evolution.* Curr Opin Plant Biol, 2010. **13**(2): p. 139-45.

374. Ito, H., et al., *Ecotype-specific and chromosome-specific expansion of variant centromeric satellites in Arabidopsis thaliana.* Mol Genet Genomics, 2007. **277**(1): p. 23-30.

375. Passamonti, M., B. Mantovani, and V. Scali, *Characterization of a highly repeated DNA family in tapetinae species (mollusca bivalvia: veneridae).* Zoolog Sci, 1998. **15**(4): p. 599-605.

376. Fadloun, A., A. Eid, and M.E. Torres-Padilla, *Mechanisms and dynamics of heterochromatin formation during Mammalian development: closed paths and open questions.* Curr Top Dev Biol, 2013. **104**: p. 1-45.

377. Wnuk, M., et al., *Changes in DNA methylation patterns and repetitive sequences in blood lymphocytes of aged horses.* Age (Dordr), 2013.

378. Brown, K.E., et al., *Atypical heterochromatin organization and replication are rapidly acquired by somatic cells following fusion-mediated reprogramming by mouse ESCs.* Cell Cycle, 2013. **12**(20): p. 3253-61.

379. Benoit, M., et al., *Heterochromatin dynamics during developmental transitions in Arabidopsis - a focus on ribosomal DNA loci.* Gene, 2013. **526**(1): p. 39-45.

380. Zhang, D., D. Wang, and F. Sun, *Drosophila melanogaster heterochromatin protein HP1b plays important roles in*

transcriptional activation and development. Chromosoma, 2011. **120**(1): p. 97-108.

381.   Ahmad, A., Y. Zhang, and X.F. Cao, *Decoding the epigenetic language of plant development.* Mol Plant, 2010. **3**(4): p. 719-28.

382.   Shvachko, L.P., *Alterations of constitutive pericentromeric heterochromatin in lymphocytes of cancer patients and lymphocytes exposed to 5-azacytidine is associated with DNA-hypomethylation.* Exp Oncol, 2008. **30**(3): p. 230-4.

383.   Gdula, M.R., et al., *Remodeling of three-dimensional organization of the nucleus during terminal keratinocyte differentiation in the epidermis.* J Invest Dermatol, 2013. **133**(9): p. 2191-201.

384.   Popova, E.Y., et al., *Chromatin condensation in terminally differentiating mouse erythroblasts does not involve special architectural proteins but depends on histone deacetylation.* Chromosome Res, 2009. **17**(1): p. 47-64.

385.   Alcobia, I., et al., *The spatial organization of centromeric heterochromatin during normal human lymphopoiesis: evidence for ontogenically determined spatial patterns.* Exp Cell Res, 2003. **290**(2): p. 358-69.

386.   Wu, B.K., S.C. Mei, and C. Brenner, *RFTS-deleted DNMT1 enhances tumorigenicity with focal hypermethylation and global hypomethylation.* Cell Cycle, 2014. **13**(20): p. 3222-31.

387.   Wu, T.F., et al., *PAX3 loads onto pericentromeric heterochromatin during S phase through PARP1.* Anticancer Res, 2014. **34**(9): p. 4717-22.

388.   Schneider, K., et al., *Dissection of cell cycle-dependent dynamics of Dnmt1 by FRAP and diffusion-coupled modeling.* Nucleic Acids Res, 2013.

389.   Ng, T.M., et al., *Pericentromeric sister chromatid cohesion promotes kinetochore biorientation.* Mol Biol Cell, 2009. **20**(17): p. 3818-27.

390.   Sakuno, T. and Y. Watanabe, *Studies of meiosis disclose distinct roles of cohesion in the core centromere and pericentromeric regions.* Chromosome Res, 2009. **17**(2): p. 239-49.

391.   Zaccarini, R., et al., *Pax6p46 binds chromosomes in the pericentromeric region and induces a mitosis defect when*

overexpressed. Invest Ophthalmol Vis Sci, 2007. **48**(12): p. 5408-
19.

392.   Wang, Q., et al., *Histone phosphorylation and pericentromeric histone modifications in oocyte meiosis.* Cell Cycle, 2006. **5**(17): p. 1974-82.

393.   Shestakova, E.A., et al., *Transcription factor YY1 associates with pericentromeric gamma-satellite DNA in cycling but not in quiescent (G0) cells.* Nucleic Acids Res, 2004. **32**(14): p. 4390-9.